Council for African American Researchers in the Mathematical Sciences: Volume V

Participants at the Thirteenth Conference for African American Researchers in the Mathematical Sciences (CAARMS 13)

CONTEMPORARY MATHEMATICS

467

Council for African American Researchers in the Mathematical Sciences: Volume V

Thirteenth Conference for African American Researchers
in the Mathematical Sciences
June 19–22, 2007
Northeastern University and the University of Massachusetts, Boston

Alfred G. Noël
Donald R. King
Gaston M. N'Guérékata
Edray H. Goins
Editors

American Mathematical Society
Providence, Rhode Island

2000 *Mathematics Subject Classification*. Primary 22Exx.

Library of Congress Cataloging-in-Publication Data

Conference for African American Researchers in the Mathematical Sciences (13th : 2007 : Boston, Mass.)

Council for African American Researchers in the Mathematical Sciences. Volume V : Thirteenth Conference for African American Researchers in the Mathematical Sciences, June 19–22, 2007, Northeastern University and the University of Massachusetts, Boston, Massachusetts / Alfred G. Noël...[et al.], editors.

p. cm. – (Contemporary mathematics ; 467)

Includes bibliographical references.

ISBN 978-0-8218-4457-1 (alk. paper)

1. Mathematics–Congresses. I. Noël, Alfred G., 1956– II. Council for African American Researchers in the Mathematical Sciences. III. Title.

QA1.C6253 2007
510—dc22 2008015382

∞ The paper used in this book is acid-free and falls within the guidelines
established to ensure permanence and durability.
Visit the AMS home page at http://www.ams.org/

10 9 8 7 6 5 4 3 2 1   13 12 11 10 09 08

In Memory
of
Fokko du Cloux
December 20, 1954 - November 10, 2006

Contents

Preface

This volume contains selected papers based upon the presentations at the Thirteenth Conference for African American Researchers in the Mathematical Sciences (CAARMS 13), held at Northeastern University and University of Massachusetts Boston, on June 19-22, 2007. The Council for African American Researchers in the Mathematical Sciences is the group that organizes this annual conference that showcases the current research primarily, but not exclusively, of African Americans in the mathematical sciences. CAARMS conferences have been held annually since 1995. At these conferences, significant numbers of researchers have delivered hour-long technical presentations and many graduate students have presented their work in organized poster sessions. The participants in past CAARMS conferences are discussed in greater detail in an article in this volume by William A. Massey, Derrick Raphael and Erica N. Walker.

Research topics featured at past CAARMS conferences have come from pure mathematics (e.g., number theory, analysis, topology, abstract algebra, and probability), mathematical physics, mathematical biology, operations research, applied statistics, and computer science. In addition to the invited talks and tutorials, group discussions are organized to stimulate, nurture, and encourage increased participation by African Americans and other under-represented groups in the mathematical sciences. These events create an ideal forum for mentoring and networking in which attendees can meet African American researchers and graduate students who are interested in the same fields.

The representation theory of Lie groups and its applications were a major focus of the talks at CAARMS 13. An article by Alfred G. Noël provides an overview of the recent achievements of the Atlas of Lie groups Project, one of the most important efforts currently underway in pure mathematics. An article by Alessandra Pantano, Annegret Paul and Susana Salamanca-Riba illustrates the work that remains to be done by the Atlas Project to understand fully the unitary representations of reductive groups. A second article by Noël highlights the significance of scientific computing in Lie theory. A paper by Floyd L. Williams applies the representation theory of Lie groups to the physics of black holes. The volume contains two research papers not directly related to Lie theory: one by Aissa Wade on the geometry of co-isotropic submanifolds of Poisson manifolds and one by Arthur D. Grainger on recent results on the structure of the set of ultrafilters on the collection of finite subsets of an infinite set.

We would like to thank the following individuals from our gracious host institutions for making the conference a success: Professor Robert C. McOwen (Northeastern University), Joanne Durham (Northeastern University), Dean emeritus Michael Greeley (University of Massachusetts, Boston), Dean William Hagar (University of Massachusetts, Boston), Provost Paul Fonteyn (University of Massachusetts, Boston) and Professor Dennis Wortman (University of Massachusetts, Boston). We thank William A. Massey for his vision, commitment and organizing efforts for this and previous CAARMS conferences. We also express our deep appreciation to all of our sponsors for their financial support: National Security Agency, Northeastern University, University of Massachusetts, Boston, Mathematical Sciences Research Institute, Clay Mathematics Institute, MITRE Corporation, Massachusetts Institute of Technology Mathematics Department, and the Webber Foundation. Finally, we gratefully acknowledge all the presenters and participants in CAARMS 13, the proceedings authors, the referees, and Christine M. Thivierge of the American Mathematical Society for supporting this conference and the preparation of this volume.

Alfred G. Noël
Donald R. King
Gaston M. N'Guérékata
Edray H. Goins

Contemporary Mathematics
Volume **467**, 2008

The Omega-Regular Unitary Dual of the Metaplectic Group of Rank 2

Alessandra Pantano, Annegret Paul, and Susana A. Salamanca-Riba

This paper is dedicated to the dear memory of Professor Fokko du Cloux.

Abstract. In this paper we formulate a conjecture about the unitary dual of the metaplectic group. We prove this conjecture for the case of Mp(4,R). The result is a strengthening, for this case, of the following result by the third author: any unitary representation of a real reductive Lie group with strongly regular infinitesimal character can be obtained by cohomological induction from a one dimensional representation. Strongly regular representations are those whose infinitesimal character is at least as regular as that of the trivial representation. We are extending the result to representations with omega-regular infinitesimal character: those whose infinitesimal character is at least as regular as that of the oscillator representation. The proof relies heavily on Parthasarathy's Dirac operator inequality. In one exception we explicitly calculate the signature of an intertwining operator to establish nonunitarity. Some of the results on intertwining operators presented in section 5.2 are joint work of Dan M. Barbasch and the first author.

1. Introduction

This paper is based on a presentation by the third author at the 13th Conference of African American Researchers in the Mathematical Sciences (CAARMS13). The presentation was intentionally expository, aimed at non-experts in the field of representation theory. With this in mind, an introductory survey of the fundamental concepts underlying this work was provided. A brief extract of the original presentation appears in the appendix. We have limited the introductory remarks to a discussion about $SL(2,\mathbb{R})$, as some results relative to this group are paramount for understanding the main ideas of the paper.

2000 *Mathematics Subject Classification.* Primary 22E46.

Key words and phrases. Unitary dual, Dirac operator inequality, intertwining operators, derived functor modules.

This material is based upon work supported by the National Science Foundation under Grants No. DMS 0554278 and DMS 0201944.

1.1. Classification of representations. Let G be a real reductive Lie group. Recall that in [**15**], Vogan gave a classification of all admissible irreducible representations of G. In fact, he gave a parametrization of all such representations containing any given irreducible representation of K as a lowest K-type. Here K is the maximal compact subgroup of G. More precisely, we have the following

PROPOSITION 1. (See [**15, 16**] for definitions and details). To a reductive Lie group G, a maximal compact subgroup K of G, and an irreducible representation μ of K, we can attach a subgroup $L_a = L_a(\mu)$ of G, a parabolic subalgebra $\mathfrak{q}_a = \mathfrak{l}_a + \mathfrak{u}_a \subseteq \mathfrak{g}$ and an $L_a \cap K$ representation μ^{L_a} such that there is a bijection

$$\mathcal{R}_{\mathfrak{q}_a}: \left\{ \begin{array}{c} (\mathfrak{l}_a, L_a \cap K) \text{ modules} \\ \text{with lowest } (L_a \cap K)\text{-type } \mu^{L_a} \end{array} \right\} \longmapsto \left\{ \begin{array}{c} (\mathfrak{g}, K) \text{ modules} \\ \text{with lowest } K\text{-type } \mu \end{array} \right\}.$$

Here $\mathfrak{g}$ and $\mathfrak{l}_a$ are the complexified Lie algebras of G and L_a, respectively. (We use similar notation for other groups and Lie algebras, and use the subscript 0 to denote real Lie algebras.)

This construction is called cohomological parabolic induction. We call $\mathcal{R}_{\mathfrak{q}}$ the cohomological induction functor. The representations on the left-hand side are minimal principal representations of the subgroup L_a. Proposition 1 essentially reduces the classification of irreducible admissible representations of G to minimal principal series of certain subgroups. If μ is the lowest K-type of a principal series representation of G then we have $L_a(\mu) = G$, and there is no reduction.

In the case of $G = SL(2,\mathbb{R})$ and $K = S^1$, $L_a(n) = K$ for $|n| \geq 2$ and $L_a(n) = G$ for $|n| \leq 1$. Here we have identified the irreducible representations of S^1 with integers in the usual way.

For unitary representations, we would like to have a statement similar to Proposition 1. In other words, we would like to have some way of classifying all the unitary representations containing a certain lowest K-type μ. This is known to be possible in some cases; in general, we have the following conjecture (see [**13**]).

CONJECTURE 1. (See [**13**]) To each representation μ of K, we can attach a subgroup L_u, a parabolic subalgebra $\mathfrak{q}_u$ and a representation μ^{L_u} of $L_u \cap K$ such that there is a bijection

$$R_{\mathfrak{q}_u}: \left\{ \begin{array}{c} \text{unitary } (\mathfrak{l}_u, L_u \cap K) \text{ modules} \\ \text{with lowest } (L_u \cap K)\text{-type } \mu^{L_u} \end{array} \right\} \longmapsto \left\{ \begin{array}{c} \text{unitary } (\mathfrak{g}, K) \text{ modules} \\ \text{with lowest } K\text{-type } \mu \end{array} \right\}.$$

In principle, μ^{L_u} is a representation for which there is no such reduction to a representation of a smaller group. However, in the case of $SL(2,\mathbb{R})$, even though we can realize the discrete series as cohomologically induced from one dimensional representations of the group T, it fits best into the general conjecture to make $L_u(n) = T$ for $|n| > 2$ and $L_u(n) = G$ for $|n| \leq 2$. This suggests that the non-reducing K-types are 0, ± 1 and ± 2. We want to have a bijection like this for any real reductive Lie group.

REMARK 1.1. As in this case, in general $\mathfrak{q}_u(\mu) \supsetneq \mathfrak{q}_a(\mu)$.

We will now provide examples of unitary representations that can be constructed from, in some sense, "smaller", or easier to understand representations of proper subgroups of G.

1.2. The $A_{\mathfrak{q}}$ representations. In this section we describe a family of unitary representations that are cohomologically induced from one-dimensional representations of a subgroup. We focus on representation that satisfy some regularity condition on the infinitesimal character. In [**12**], the strongly regular case was considered (see 1.2.1). In the present paper we consider a weakening of the regularity assumption for representations of the metaplectic groups $Mp(2n,\mathbb{R})$, $n=1,2$.

1.2.1. *Strongly regular case.* Let G be reductive. Let $\mathfrak{h}=\mathfrak{t}+\mathfrak{a}$ be a maximally compact Cartan subalgebra of $\mathfrak{g}$, with $\mathfrak{t}$ a Cartan subalgebra of $\mathfrak{k}$. For a weight $\phi\in\mathfrak{h}^*$, choose a positive root system from the set of roots positive on ϕ:

$$\Delta^+(\phi)\subseteq\{\alpha\in\Delta(\mathfrak{g},\mathfrak{t})\mid\langle\phi,\alpha\rangle\geq 0\}.$$

Then define

$$\rho_\phi=\rho\left(\Delta^+(\phi)\right)=\frac{1}{2}\sum_{\alpha\in\Delta^+(\phi)}\alpha.$$

DEFINITION 1.2. Suppose $\phi\in\mathfrak{h}^*$ is real. We say that ϕ is strongly regular if $\langle\phi-\rho_\phi,\alpha\rangle\geq 0$ for all $\alpha\in\Delta^+(\phi)$.

1.2.2. $A_{\mathfrak{q}}(\lambda)$ *representations.* Recall that a theta stable parabolic subalgebra $\mathfrak{q}=\mathfrak{l}+\mathfrak{u}$ of $\mathfrak{g}$ is defined as the sum of the nonnegative root spaces for $ad(\xi)$ where ξ is an element of $i\mathfrak{t}_0$. The Levi subalgebra $\mathfrak{l}$ is the zero eigenspace and contains $\mathfrak{t}$. It is a reductive subalgebra of $\mathfrak{g}$. The sum of the positive eigenspaces is the nilradical $\mathfrak{u}$ of $\mathfrak{q}$. Let L be the Levi subgroup of G corresponding to $\mathfrak{l}$. Then $\mathfrak{l}_0$ is the Lie algebra of L. We construct a representation of G as follows.

DEFINITION 1.3. For every one-dimensional representation $\mathbb{C}_\lambda$ of L satisfying

$$\langle\lambda|_{\mathfrak{t}},\alpha\rangle\geq 0\qquad\forall\alpha\in\Delta(\mathfrak{u})\tag{1.1}$$

we define $A_{\mathfrak{q}}(\lambda):=R_{\mathfrak{q}}(\mathbb{C}_\lambda)$.

Here $\Delta(\mathfrak{u})=\Delta(\mathfrak{u},\mathfrak{t})$. In general, for any $\mathfrak{t}$-invariant subspace $\mathfrak{s}\subseteq\mathfrak{g}$, we write $\Delta(\mathfrak{s})=\Delta(\mathfrak{s},\mathfrak{t})$ for the set of weights of $\mathfrak{t}$ in $\mathfrak{s}$ counted with multiplicities.

REMARK 1.4. All $A_{\mathfrak{q}}(\lambda)$ representations constructed this way are nonzero, irreducible and unitary.

PROPOSITION 2. [**12**] Suppose G is a real reductive Lie group and X is an irreducible Hermitian $(\mathfrak{g},K)$ module with a real, strongly regular infinitesimal character. Then X is unitary if and only if there is a parabolic subalgebra $\mathfrak{q}$ of $\mathfrak{g}$ and a one-dimensional representation $\mathbb{C}_\lambda$ of L satisfying (1.1) and such that

$$X\simeq A_{\mathfrak{q}}(\mathbb{C}_\lambda).$$

1.2.3. *The omega-regular case and the* $A_{\mathfrak{q}}(\Omega)$ *representations of* $Mp(2n)$. Let $G=Mp(2n)$, the connected double cover of the group $Sp(2n,\mathbb{R})$. Then $\mathfrak{g}=\mathfrak{sp}(2n)$. Representations of G may be divided into those which factor through $Sp(2n,\mathbb{R})$ ("nongenuine" ones), and those that do not, the "genuine" representations. The nongenuine representations of G are essentially the representations of the linear group; in particular, a nongenuine representation of G is unitary if and only if the corresponding representation of $Sp(2n,\mathbb{R})$ is.

In order to build a genuine $A_{\mathfrak{q}}(\lambda)$ representation of G, we need to start with a genuine one-dimensional representation of the Levi subgroup L corresponding to the Levi factor $\mathfrak{l}$ of a theta stable parabolic subalgebra of $\mathfrak{g}$. Such subgroups are (quotients of) products of factors isomorphic to smaller metaplectic groups and double covers of $U(p,q)$'s. Notice that the metaplectic group does not have any genuine one-dimensional representation, hence there are no $A_{\mathfrak{q}}(\lambda)$ representations for such $\mathfrak{q}$. We extend our definition to allow the oscillator representation ω, a minimal genuine unitary representation of $Mp(2m)$ on such factors. The infinitesimal character of the oscillator representation is not strongly regular, but satisfies a slightly weaker condition, which we call "omega-regular" (see Definition 2.1 for a precise definition). If we apply cohomological induction to representations of L of the form

$$\Omega = \mathbb{C}_\lambda \otimes \omega \tag{1.2}$$

then, by a construction analogous to the one of the $A_{\mathfrak{q}}(\lambda)$ representations, we obtain genuine irreducible unitary representations of G with ω-regular infinitesimal character, which we call $A_{\mathfrak{q}}(\Omega)$ representations (see Definition 2.4).

If we hope to list all unitary ω-regular representations of G, we must extend our definition of $A_{\mathfrak{q}}(\Omega)$ representations to the nongenuine case as well, since the representations $I_P(\delta_+, u)$ of $SL(2,\mathbb{R})$ with $\frac{1}{2} < u < 1$ ("complementary series") in Table 1 (6.8) are ω-regular and unitary, but not $A_{\mathfrak{q}}(\lambda)$ modules. We define a family of nongenuine ω-regular unitary representations of G, which we call Meta-$A_{\mathfrak{q}}(\lambda)$ representations, by allowing complementary series on any $Mp(2)$ factor of L, and relaxing condition 1.1 somewhat (see Definition 2.5).

CONJECTURE 2. Let G be $Mp\,(2n)$ and let X be a genuine irreducible representation of G with real infinitesimal character. Then X is ω-regular and unitary if and only if there are $\mathfrak{q}$, L and a genuine representation Ω of L as above such that

$$X \simeq A_{\mathfrak{q}}\,(\Omega)\,.$$

If X is nongenuine, then X is ω-regular and unitary if and only if X is a Meta-$A_{\mathfrak{q}}\,(\lambda)$ representation.

The main result of this paper is a proof of the conjecture for $Mp\,(2)$ and $Mp\,(4)$ (see Conjecture 3 and Theorem 2.8). The case of $Mp\,(2n)$ with $n \geq 3$ has many additional interesting and complicating features and will appear in a future paper.

The full unitary duals of $Mp(2)$ and $Sp(4,\mathbb{R})$ are well known (c.f. [**4**] and [**10**]); some basic results are reported for the sake of completeness. The most innovative part of this paper regards genuine representations of $Mp(4)$. The proof of the conjecture in this case requires more elaborate techniques. A synopsis follows. First, we determine the set of genuine representations of K which are lowest K-types of $A_{\mathfrak{q}}\,(\Omega)$ representations; for every representation of K in this list, we establish that there is a unique unitary irreducible representation of $Mp(4)$ with that lowest K-type. Then, we consider genuine representations of K which are *not* lowest K-types of $A_{\mathfrak{q}}\,(\Omega)$ representations, and we establish that any ω-regular representation of $Mp(4)$ containing those K-types is nonunitary. It turns out that Parthasarathy's Dirac operator inequality can be used to prove nonunitary for all but two representations. The last section of the paper is dedicated to proving that these two remaining K-types cannot occur in any unitary representation. The proof is based on an explicit calculation of the signature of the intertwining operator. Some results about intertwining operators are included in Section 5.2.

The paper is organized as follows. In Section 2, we define our notation and state some preliminary facts and results. Sections 3, 4 and 5 contain the proof of

Conjecture 2 for $Mp\,(2)$ and $Mp\,(4)$. Section 3 is dedicated to $Mp(2)$, the remaining sections deal with $Mp(4)$.

The authors thank Jeffrey Adams and David Vogan for posing the problem, and for their help and support along the way. They would also like to thank Dan M. Barbasch for his invaluable help in developing the theory of intertwining operators, and for his generosity with both his time and his ideas. The third author wishes to thank Donald R. King, Alfred G. Noel and William Massey, organizers of the CAARMS13 conference, for their invitation to speak at the conference and to submit this paper to these Proceedings.

2. Preliminaries

2.1. Setup. Let $G = Mp(2n) = Mp(2n,\mathbb{R})$ be the metaplectic group, i. e., the connected double cover of the symplectic group $Sp(2n,\mathbb{R})$, and denote by

$$pr\colon Mp(2n) \to Sp(2n,\mathbb{R}) \tag{2.1}$$

the covering map. Fix a Cartan decomposition $\mathfrak{g} = \mathfrak{k} + \mathfrak{p}$ of $\mathfrak{g} = \mathfrak{sp}(2n,\mathbb{C})$, and let θ be the corresponding Cartan involution. Let $\mathfrak{q} = \mathfrak{l} + \mathfrak{u}$ be a theta stable parabolic subalgebra of $\mathfrak{g} = \mathfrak{sp}(2n,\mathbb{C})$. Then the Levi subgroup L of $Mp(2n)$ corresponding to $\mathfrak{l}$ is the inverse image under pr of a Levi subgroup of $Sp(2n,\mathbb{R})$ of the form

$$\prod_{i=1}^{r} U(p_i,q_i) \times Sp(2m,\mathbb{R}). \tag{2.2}$$

There is a surjection

$$\prod_{i=1}^{r} \widetilde{U}(p_i,q_i) \times Mp(2m) \to L, \tag{2.3}$$

where $\widetilde{U}(p_i,q_i)$ denotes the connected "square root of the determinant" cover of $U(p_i,q_i)$,

$$\widetilde{U}(p_i,q_i) \simeq \left\{(g,z) \in U(p_i,q_i) \times \mathbb{C}^{\times} : z^2 = \det(g)\right\}. \tag{2.4}$$

An irreducible admissible representation of L may be given by a representation

$$\bigotimes_{i=1}^{r} \pi_i \otimes \sigma, \tag{2.5}$$

where π_i is an irreducible admissible representation of $\widetilde{U}(p_i,q_i)$ for each i, and σ is an irreducible admissible representation of $Mp(2m)$. In order for this tensor product to descend to a representation of L, we must have that either all representations in the product are genuine, i. e., nontrivial on the kernel of the covering map, or all representations are nongenuine. In the first case, the representation σ of $Mp(2m)$ will then be genuine. In the second case, it will factor through $Sp(2m,\mathbb{R})$. With this in mind, we will often identify L with the product in (2.3), and a representation of L with a representation of the product.

In most cases, the representation $\pi = \bigotimes_{i=1}^{r} \pi_i$ we consider will be one-dimensional, and we denote it by $\mathbb{C}_\lambda$. The genuine representations of $Mp(2m)$ we consider will be the four oscillator representations $\omega_o^{\pm}$, $\omega_e^{\pm}$. Here $\omega^+ = \omega_e^+ + \omega_o^+$ denotes the holomorphic oscillator representation which is a sum of the even and odd constituents, and $\omega^- = \omega_e^- + \omega_o^-$ is its contragredient, the antiholomorphic oscillator representation of $Mp(2m)$. We will often refer to any of these four irreducible representations as "an oscillator representation of $Mp(2m)$".

The nongenuine representations of $Mp(2m)$ will be the trivial representation $\mathbb{C}$ or, in the case of $Mp(2)$, the unique spherical constituents of the spherical principal series representations J_ν with infinitesimal character ν satisfying $\frac{1}{2} \leq \nu \leq 1$ (the "complementary series representations"). Recall (see Table 1) that these representations are unitary, and J_1 is the trivial representation of $Mp(2)$.

Let $\mathfrak{t}$ be a fundamental Cartan subalgebra of $\mathfrak{g}$. Recall that $\mathfrak{t}$ is also a Cartan subalgebra for $\mathfrak{k}$, the complexified Lie algebra of $K \simeq \widetilde{U}(n)$, a maximal compact subgroup of G. Let $\Delta(\mathfrak{g},\mathfrak{t}) \subseteq i\mathfrak{t}_0^*$ be the set of roots of $\mathfrak{t}$ in $\mathfrak{g}$. (Here, as everywhere else in the paper, we use the subscript 0 to denote real Lie algebras.) With respect to a standard parametrization, we can identify elements of $i\mathfrak{t}_0^*$ with n-tuples of real numbers. With this identification,

$$\Delta(\mathfrak{g},\mathfrak{t}) = \{\pm 2e_i : 1 \leq i \leq n\} \cup \{\pm e_i \pm e_j : 1 \leq i < j \leq n\}, \tag{2.6}$$

where e_i is the n-tuple with 1 in the ith position, and 0 everywhere else. Then the compact roots are

$$\Delta_k = \Delta(\mathfrak{k},\mathfrak{t}) = \{\pm(e_i - e_j) : 1 \leq i < j \leq n\}. \tag{2.7}$$

We fix a system of positive compact roots

$$\Delta_k^+ = \{e_i - e_j : 1 \leq i < j \leq n\} \tag{2.8}$$

and write

$$\rho_c = \left(\frac{n-1}{2}, \frac{n-3}{2}, \dots, \frac{-n+1}{2}\right), \tag{2.9}$$

one half the sum of the roots in Δ_k^+. We identify K-types, i. e., irreducible representations of K, with their highest weights which will be given by n-tuples of weakly decreasing integers (if nongenuine) or elements of $\mathbb{Z} + \frac{1}{2}$ (if genuine). The lowest K-types (in the sense of Vogan [**16**]) of $\omega_e^+, \omega_o^+, \omega_e^-$, and ω_o^- are

$$\left(\frac{1}{2}, \frac{1}{2}, \dots, \frac{1}{2}\right), \tag{2.10}$$

$$\left(\frac{3}{2}, \frac{1}{2}, \frac{1}{2}, \dots, \frac{1}{2}\right), \tag{2.11}$$

$$\left(-\frac{1}{2}, -\frac{1}{2}, \dots, -\frac{1}{2}\right), \text{ and} \tag{2.12}$$

$$\left(-\frac{1}{2}, -\frac{1}{2}, \dots, -\frac{1}{2}, -\frac{3}{2}\right) \tag{2.13}$$

respectively.

Using the Harish-Chandra map, we identify infinitesimal characters of admissible representations of G with (Weyl group orbits of) elements of $\mathfrak{t}^*$. Recall that the Weyl group $W(\mathfrak{g},\mathfrak{t})$ acts on $\mathfrak{t}$ by permutations and sign changes. For example, the infinitesimal character γ_ω of any of the oscillator representations can be represented by the element

$$\left(n - \frac{1}{2}, n - \frac{3}{2}, \dots, \frac{3}{2}, \frac{1}{2}\right); \tag{2.14}$$

we will often abuse notation by writing

$$\gamma_\omega = \left(n - \frac{1}{2}, n - \frac{3}{2}, \ldots, \frac{3}{2}, \frac{1}{2}\right). \tag{2.15}$$

We fix a non-degenerate G- and θ-invariant symmetric bilinear form $< \,,\, >$ on $\mathfrak{g}_0$, and we use the same notation for its various restrictions, extensions, and dualizations. In our parametrization of elements of $\mathfrak{t}$, this is the standard inner product

$$((a_1, \ldots, a_n), (b_1, \ldots, b_n)) = \sum_{i=1}^{n} a_i b_i. \tag{2.16}$$

2.2. Definitions and Conjecture.

DEFINITION 2.1. Let $\gamma \in i\mathfrak{t}_0^*$. Choose a positive system $\Delta^+(\gamma) \subseteq \Delta(\mathfrak{g}, \mathfrak{t})$ such that $\langle \alpha, \gamma \rangle \geq 0$ for all $\alpha \in \Delta^+(\gamma)$, and let γ_ω be the representative of the infinitesimal character of the oscillator representation which is dominant with respect to $\Delta^+(\gamma)$. We call γ *ω-regular* if the following regularity condition is satisfied:

$$\langle \alpha, \gamma - \gamma_\omega \rangle \geq 0 \quad \forall \alpha \in \Delta^+(\gamma). \tag{2.17}$$

We say that a representation of G is ω-regular if its infinitesimal character is.

REMARK 2.2. The definition of ω-regular infinitesimal character is similar to the one of strongly regular infinitesimal character (c.f. [**12**]), but uses the infinitesimal character γ_ω of the oscillator representation instead of the infinitesimal character ρ of the trivial representation. Note that every strongly regular infinitesimal character is necessarily ω-regular. We will prove this result in the course of the proof of proposition 3.

EXAMPLE 2.3. In $Mp(2)$, an infinitesimal character $\gamma = (k)$ is ω-regular if k is a real number such that $|k| \geq \frac{1}{2}$; it is strongly regular if $|k| \geq 1$.
In $Mp(4)$, an infinitesimal character $\gamma = (a, b)$ is ω-regular if a and b are both real, $||a| - |b|| \geq 1$, and $\min\{|a|, |b|\} \geq \frac{1}{2}$. It is strongly regular if, in addition, $\min\{|a|, |b|\} \geq 1$.

We will focus on two families of ω-regular representations: the $A_\mathfrak{q}(\Omega)$ and the *Meta-$A_\mathfrak{q}(\lambda)$* representations of G, which we define below. In both cases, $\mathfrak{q} = \mathfrak{l} + \mathfrak{q}$ is a theta stable parabolic subalgebra of $\mathfrak{g}$ with

$$L = \prod_{i=1}^{r} \widetilde{U}(p_i, q_i) \times Mp(2m).$$

We write $\rho(\mathfrak{u})$ for one half the sum of the roots of $\mathfrak{u}$.

DEFINITION 2.4. An $A_\mathfrak{q}(\Omega)$ representation is a genuine representation of G of the following form. Let $\mathbb{C}_\lambda$ be a genuine one-dimensional representation of $\left[\prod_{i=1}^{r} \widetilde{U}(p_i, q_i)\right]$ and let ω be an oscillator representation of $Mp(2m)$. Assume that the representation $\Omega = \mathbb{C}_\lambda \otimes \omega$ of L is in the good range for $\mathfrak{q}$, i. e., that the infinitesimal character γ^L of Ω is such that $\gamma^L + \rho(\mathfrak{u})$ is strictly dominant with respect to the roots of $\mathfrak{u}$. We define

$$A_\mathfrak{q}(\Omega) := \mathcal{R}_\mathfrak{q}(\Omega).$$

$\mathcal{R}_\mathfrak{q}$ denotes the right cohomological induction functor defined in [**15**] and [**16**].

DEFINITION 2.5. A *Meta-*$A_{\mathfrak{q}}(\lambda)$ representation is a nongenuine representation X of G of the following form. Let $\mathbb{C}_\lambda$ be a nongenuine one-dimensional representation of $\left[\prod_{i=1}^{r} \widetilde{U}(p_i, q_i)\right]$ and let J_ν be the spherical constituent of the spherical principal series of $Mp(2m)$ with infinitesimal character ν. If $m \neq 1$ then we take $\nu = \rho$ so that $J_\nu = J_\rho$ is the trivial representation of $Mp(2m)$; if $m = 1$ then require that $\frac{1}{2} \leq \nu \leq 1$, so that J_ν is a complementary series of $Mp(2)$ if $\frac{1}{2} \leq \nu < 1$ and is the trivial representation if $\nu = 1$. Assume that $\mathbb{C}_\lambda \otimes J_\nu$ is in the good range for $\mathfrak{q}$. We define

$$X := \mathcal{R}_{\mathfrak{q}}(\mathbb{C}_\lambda \otimes J_\nu).$$

REMARK 2.6. Every $A_{\mathfrak{q}}(\lambda)$ representation of $Mp(2n)$ in the good range is either an $A_{\mathfrak{q}}(\Omega)$ or a Meta-$A_{\mathfrak{q}}(\lambda)$ representation.
More explicitly, if X is a genuine $A_{\mathfrak{q}}(\lambda)$ representation in the good range, then we can consider X as an $A_{\mathfrak{q}}(\Omega)$ representation with $m = 0$ (note that, in this case, the Levi subgroup L does not contain any $Sp(2m)$ factor). If X is a nongenuine $A_{\mathfrak{q}}(\lambda)$ representation in the good range, then we can consider X as a Meta-$A_{\mathfrak{q}}(\lambda)$ representation with J_ν equal to the trivial representation of $Sp(2m)$ (for all m).

PROPOSITION 3. The following properties hold:

(1) All $A_{\mathfrak{q}}(\Omega)$ and Meta-$A_{\mathfrak{q}}(\lambda)$ representations of G are nonzero, irreducible and unitary.
(2) If X is a Meta-$A_{\mathfrak{q}}(\lambda)$ representation with $\nu = \rho$, then X is an admissible $A_{\mathfrak{q}}(\lambda)$ in the sense of [**12**], and has strongly regular infinitesimal character.
(3) All Meta-$A_{\mathfrak{q}}(\lambda)$ representations of G are ω-regular.
(4) All $A_{\mathfrak{q}}(\Omega)$ representations of G are ω-regular.

The proof of this Proposition will be given at the end of this section.

REMARK 2.7. Genuine $A_{\mathfrak{q}}(\lambda)$ representations in the good range are not necessarily strongly regular. For example, take $G = Mp(4)$, $\mathfrak{q} = \mathfrak{l} + \mathfrak{u}$ with $L \cong \widetilde{U}(1,1)$, so that $\rho(\mathfrak{u}) = \left(\frac{3}{2}, -\frac{3}{2}\right)$ and $\rho(\mathfrak{l}) = \left(\frac{1}{2}, \frac{1}{2}\right)$, and choose $\lambda = \left(-\frac{1}{2}, \frac{1}{2}\right)$. The module $A_{\mathfrak{q}}(\lambda)$ has lowest K-type

$$\mu = \lambda + 2\rho(\mathfrak{u} \cap \mathfrak{p}) = \left(-\frac{1}{2}, \frac{1}{2}\right) + (2, -2) = \left(\frac{3}{2}, -\frac{3}{2}\right) \tag{2.18}$$

and infinitesimal character

$$\gamma = \lambda + \rho(\mathfrak{l}) + \rho(\mathfrak{u}) = \left(\frac{3}{2}, -\frac{1}{2}\right) = \gamma_\omega. \tag{2.19}$$

Now we are ready to state our conjecture.

CONJECTURE 3. Let X be an irreducible admissible representation of $Mp(2n)$. Then X is ω-regular and unitary if and only if X is either an $A_{\mathfrak{q}}(\Omega)$ or a Meta-$A_{\mathfrak{q}}(\lambda)$.

THEOREM 2.8. *Conjecture 3 is true for* $n = 1$ *and* $n = 2$.

The proof of Theorem 2.8 will occupy most of the remainder of this paper. Before proving Proposition 3 (and other facts about $A_{\mathfrak{q}}(\Omega)$ and Meta-$A_{\mathfrak{q}}(\lambda)$ representations), we need to collect a few results on cohomological parabolic induction.

Fix a parabolic subalgebra $\mathfrak{q} = \mathfrak{l} + \mathfrak{u} \subseteq \mathfrak{g}$, and a (Levi) subgroup $L = N_G(\mathfrak{q})$. The cohomological parabolic induction functor $\mathfrak{R}_{\mathfrak{q}}$, defined in [**16**, Def. 6.3.1], maps

$(\mathfrak{l}, L\cap K)$ modules to $(\mathfrak{g}, K)$ modules. Its restriction to K, denoted by $\mathfrak{R}_{\mathfrak{q}}^K$, maps $(L\cap K)$-modules to K-modules.

PROPOSITION 4. ([**17**, Lemma 6.5]). Let W be an irreducible representation of $L\cap K$, and let μ^L be a highest weight of W. Set

$$\mu = \mu^L + 2\rho\,(\mathfrak{u}\cap\mathfrak{p})\,.$$

If μ is dominant for K, then every irreducible constituent of $\mathfrak{R}_{\mathfrak{q}}^K\,(W)$ has highest weight μ. Otherwise, $\mathfrak{R}_{\mathfrak{q}}^K\,(W) = 0$.

PROPOSITION 5. ([**17**, Theorems 1.2, 1.3] and [**6**, Theorem 10.44]). Suppose that the group $L = N_G\,(\mathfrak{q})$ meets every component of G and that $\mathfrak{h}\subseteq\mathfrak{l}$ is a Cartan subalgebra. Let Y be an $(\mathfrak{l}, L\cap K)$ module, and let $\gamma^L\in\mathfrak{h}^*$ be a weight associated to the infinitesimal character of Y. Then

(1) The weight $\gamma = \gamma^L + \rho\,(\mathfrak{u})\in\mathfrak{h}^*$ is attached to the infinitesimal character of the representation $\mathfrak{R}_{\mathfrak{q}}Y$.
(2) If Y is in the good range for $\mathfrak{q}$, that is

$$\operatorname{Re}\left\langle\gamma^L + \rho\,(\mathfrak{u})\,,\alpha\right\rangle > 0 \quad \forall\alpha\in\Delta(\mathfrak{u}), \tag{2.20}$$

the following additional properties hold:
   (a) If Y is irreducible and unitary, then $\mathfrak{R}_{\mathfrak{q}}Y$ is irreducible, non zero and unitary.
   (b) The correspondence

$$\delta^L \longmapsto \delta = \delta^L + 2\rho\,(\mathfrak{u}\cap\mathfrak{p})$$

   gives a bijection between lowest $(L\cap K)$-types of Y and lowest K-types of $\mathfrak{R}_{\mathfrak{q}}Y$. In fact, every such expression for δ is dominant for K.

REMARK 2.9. If the inequality in equation 2.20 is not strict, then the induced module $\mathfrak{R}_{\mathfrak{q}}Y$ may be zero or not unitary, and some of the lowest $(L\cap K)$-types δ^L may give rise to weights for K that are not dominant.

We now give the proof of Proposition 3.

PROOF. Part (1) of Proposition 3 follows directly from Proposition 5, because both Ω and $\mathbb{C}_\lambda\otimes J_\nu$ are assumed to be in the good range for $\mathfrak{q}$.

For the second part, write γ^L for the infinitesimal character of $Z = \mathbb{C}_\lambda\otimes J_\nu$. Assume Z is in the good range for $\mathfrak{q}$, so that $\gamma^L + \rho\,(\mathfrak{u})$ is strictly dominant for the roots of $\Delta(\mathfrak{u})$, and choose $\nu = \rho$. Note that Z has infinitesimal character

$$\gamma^L = \lambda + \rho(\mathfrak{l})$$

for some choice of positive roots $\Delta^+(\mathfrak{l})\subset\Delta(\mathfrak{l},\mathfrak{t})$. By Proposition 5, the infinitesimal character of $X = \mathfrak{R}_{\mathfrak{q}}Z$ is

$$\gamma = \gamma^L + \rho\,(\mathfrak{u}) = \lambda + \rho(\mathfrak{l}) + \rho\,(\mathfrak{u}) = \lambda + \rho.$$

Here ρ is one half the sum of the roots in $\Delta^+(\mathfrak{g}) = \Delta^+(\mathfrak{l})\cup\Delta\,(\mathfrak{u})$. We want to prove that γ is strongly regular.
If $\alpha\in\Delta^+\,(\mathfrak{l})$, then

$$\langle\gamma-\rho,\alpha\rangle = \langle\lambda,\alpha\rangle = 0 \tag{2.21}$$

because λ is the differential of a one-dimensional representation of L.
If α is a simple root in $\Delta(\mathfrak{u})$, then $\langle \rho, \alpha \rangle = 1$ or 2, depending on whether α is short or long, and $\langle \gamma, \alpha \rangle > 0$ (by the "good range" condition). So

$$\begin{cases} \langle \gamma - \rho, \alpha \rangle > -1 & \text{if } \alpha \text{ is short} \\ \langle \gamma - \rho, \alpha \rangle > -2 & \text{if } \alpha \text{ is long.} \end{cases}$$

Now, because λ is the differential of a nongenuine character, the inner product $\langle \lambda, \alpha \rangle$ has integer values for all roots in $\Delta^+(\mathfrak{g})$:

$$\langle \gamma - \rho, \alpha \rangle = \langle \lambda, \alpha \rangle \in \mathbb{Z} \quad \forall \alpha \in \Delta^+(\mathfrak{g}).$$

Notice that $\langle \lambda, \alpha \rangle$ is an even integer if α is long. Then

$$\langle \gamma - \rho, \alpha \rangle = \langle \lambda, \alpha \rangle \geq 0 \tag{2.22}$$

for every (simple) root in $\Delta^+(\mathfrak{u})$. Combining this result with equation 2.21 we find that

$$\langle \gamma - \rho, \alpha \rangle = \langle \lambda, \alpha \rangle \geq 0 \quad \forall \alpha \in \Delta^+(\mathfrak{g}) = \Delta^+(\mathfrak{l}) \cup \Delta(\mathfrak{u}). \tag{2.23}$$

Hence our Meta-$A_{\mathfrak{q}}(\lambda)$ representation is admissible in the sense of [**12**] (c.f. (1.1)). Equation 2.23 also shows that $\gamma - \rho$ (and therefore γ) is weakly dominant with respect to the positive root system $\Delta^+(\mathfrak{g})$, so γ lies in the Weyl chamber of ρ, and we can take $\Delta^+(\gamma) = \Delta^+(\mathfrak{g})$ (c.f. Definition 2.1 and the remark following it). We conclude that γ is strongly regular.

For the third part of Proposition 3, we must show that every Meta-$A_{\mathfrak{q}}(\lambda)$ representation is ω-regular. If $\nu = \rho$, this result is not hard to prove. Indeed, strong regularity easily implies ω-regularity: assume that γ is strongly regular and let γ_ω be the representative of the infinitesimal character of the oscillator representation which is in the Weyl chamber of ρ. Then for every simple root $\alpha \in \Delta^+(\mathfrak{g})$, we have

$$\langle \gamma_\omega, \alpha \rangle = 1. \tag{2.24}$$

In particular, $\langle \gamma_\omega, \alpha \rangle \leq \langle \rho, \alpha \rangle$, so we get

$$\langle \gamma - \gamma_\omega, \alpha \rangle \geq \langle \gamma - \rho, \alpha \rangle \geq 0,$$

proving that γ is omega-regular.
Now assume that $\nu \neq \rho$. Then J_ν is the irreducible quotient of a complementary series representation of $Mp(2)$ with $\frac{1}{2} \leq \nu < 1$. The infinitesimal character of J_ν is equal to ν, and the restriction of $\lambda + \rho(\mathfrak{l})$ to $\mathfrak{t} \cap \mathfrak{sp}(2)$ is 1, so the infinitesimal characters of Z and $X = \mathcal{R}_{\mathfrak{q}}(Z)$ can be written as

$$\gamma^L = \lambda + \rho(\mathfrak{l}) + (\nu - 1), \tag{2.25}$$

and

$$\gamma = \lambda + \rho + (\nu - 1) \tag{2.26}$$

respectively. Assume that γ^L is in the good range, and note that because $\frac{1}{2} \leq \nu < 1$ and λ is integral, this condition is equivalent to requiring that $\lambda + \rho(\mathfrak{l})$ be in the good range. The same argument used in the second part of the proof shows that λ is weakly dominant with respect to $\Delta^+(\mathfrak{g})$. Then for all simple roots δ we have $\langle \rho, \delta \rangle \geq 1$, $\langle (\nu - 1), \delta \rangle \geq -1$ and

$$\langle \gamma, \delta \rangle = \langle \lambda + \rho + (\nu - 1), \delta \rangle \geq 0 + 1 - 1 = 0.$$

This proves that γ lies in the Weyl chamber determined by ρ, so we can take $\Delta^+(\gamma) = \Delta^+(\mathfrak{g})$, and γ is ω-regular if and only if

$$\langle \gamma - \gamma_\omega,\, \alpha \rangle \geq 0 \quad \forall \alpha \in \Delta^+(\mathfrak{g}).$$

It is sufficient to restrict the attention to the simple roots that are not orthogonal to $(\nu - 1)$: if $\langle (\nu-1),\, \alpha \rangle = 0$, then the proof for the previous case ($\nu = 1$) goes through, so $\langle \gamma - \gamma_\omega,\, \alpha \rangle \geq 0$ by the previous argument.
There are two simple roots in $\Delta^+(\mathfrak{g})$ not orthogonal to $(\nu - 1)$: a long root $\beta_L \in \Delta^+(\mathfrak{l})$ satisfying

$$\langle \nu - 1,\, \beta_L \rangle = 2\nu - 2,$$

and a short root $\beta_S \in \Delta(\mathfrak{u})$ satisfying

$$\langle \nu - 1,\, \beta_S \rangle = 1 - \nu.$$

Because β_L is simple and long, $\langle \rho, \beta_L \rangle = 2$ and $\langle \gamma_\omega, \beta_L \rangle = 1$ (from 2.24). Then

$$\begin{aligned}\langle \gamma - \gamma_\omega,\, \beta_L \rangle &= \langle \lambda + (\rho - \gamma_\omega) + (\nu - 1),\, \beta_L \rangle = \\ &= \langle \lambda, \beta_L \rangle + \langle \rho - \gamma_\omega, \beta_L \rangle + \langle (\nu - 1), \beta_L \rangle = \\ &= 0 + (2 - 1) + (2\nu - 2) = 2\nu - 1 \geq 0.\end{aligned}$$

Similarly, because β_S is simple and short, we have $\langle \rho - \gamma_\omega, \beta_S \rangle = 1 - 1 = 0$, and

$$\langle \gamma - \gamma_\omega,\, \beta_S \rangle = \langle \lambda,\, \beta_S \rangle + \langle (\nu - 1),\, \beta_S \rangle = \langle \lambda,\, \beta_S \rangle + (1 - \nu) > \langle \lambda,\, \beta_S \rangle \geq 0.$$

This proves that γ is ω-regular.

Finally, we prove part 4 of Proposition 3. Assume that $\Omega = \mathbb{C}_\lambda \otimes \omega$ is in the good range for $\mathfrak{q}$. Recall that ω is an oscillator representation of $Mp(2m) \subset L$. Set $\mathfrak{l}_1 = \mathfrak{sp}\,(2m, \mathbb{C}) \subset \mathfrak{l}$ and

$$\Lambda_{\mathfrak{l}_1} = \left\{ \beta \in \Delta^+\,(\mathfrak{l}_1, \mathfrak{t}) : \beta \text{ is long} \right\}.$$

Then

- ω has infinitesimal character $\gamma_\omega^{\mathfrak{l}_1} = \rho(\mathfrak{l}_1) - \frac{1}{2}\rho\,(\Lambda_{\mathfrak{l}_1})$
- Ω has infinitesimal character $\gamma^L = \lambda + \rho(\mathfrak{l}) - \frac{1}{2}\rho\,(\Lambda_{\mathfrak{l}_1})$
- $A_\mathfrak{q}(\Omega)$ has infinitesimal character $\gamma = \gamma^L + \rho(\mathfrak{u}) = \lambda + \rho - \frac{1}{2}\rho\,(\Lambda_{\mathfrak{l}_1})$.

We need to show that γ is ω-regular, i.e.

$$\langle \gamma - \gamma_\omega^{\mathfrak{g}},\, \alpha \rangle \geq 0 \quad \forall \alpha \in \Delta^+(\gamma). \tag{2.27}$$

Here $\gamma_\omega^{\mathfrak{g}}$ is an infinitesimal character for an oscillator representation of G. We can write

$$\gamma_\omega^{\mathfrak{g}} = \rho - \frac{1}{2}\rho\,(\Lambda_\mathfrak{g})$$

with

$$\Lambda_\mathfrak{g} = \left\{ \beta \in \Delta^+\,(\mathfrak{g}, \mathfrak{t}) : \beta \text{ is long} \right\}.$$

So equation 2.27 is equivalent to:

$$\left\langle \lambda + \frac{1}{2}\rho\,(\Lambda_\mathfrak{g}) - \frac{1}{2}\rho\,(\Lambda_1),\, \alpha \right\rangle \geq 0 \quad \forall \alpha \in \Delta^+(\mathfrak{g}). \tag{2.28}$$

Choose $w \in W(\mathfrak{g}, \mathfrak{t})$ such that $w\Delta^+(\mathfrak{g})$ is the standard positive system of roots:

$$w\Delta^+(\mathfrak{g}) = \{2e_i \colon 1 \leq i \leq n\} \cup \{e_i \pm e_j \colon 1 \leq i < j \leq n\}. \tag{2.29}$$

Then

$$w\lambda = \Big(\underbrace{\lambda_1,\dots,\lambda_1}_{p_1+q_1},\underbrace{\lambda_2,\dots,\lambda_2}_{p_2+q_2},\dots,\underbrace{\lambda_r,\dots,\lambda_r}_{p_r+q_r},\underbrace{0,\dots,0}_{m}\Big) \tag{2.30}$$

for some $\lambda_i \in \mathbb{Z}+\frac{1}{2}$ (since $\mathbb{C}_\lambda$ is genuine), and

$$w\gamma = \Big(\underbrace{\lambda_1,\dots,\lambda_1}_{p_1+q_1},\underbrace{\lambda_2,\dots,\lambda_2}_{p_2+q_2},\dots,\underbrace{\lambda_r,\dots,\lambda_r}_{p_r+q_r},\underbrace{-\frac{1}{2},\dots,-\frac{1}{2}}_{m}\Big) + w\rho \tag{2.31}$$

with $w\rho = (n, n-1, \dots, 2, 1)$. By assumption, Ω is in the good range for $\mathfrak{q}$, so γ is strictly dominant for the roots of $\mathfrak{u}$. We also have

$$\langle w\gamma, w\alpha\rangle > 0 \quad \forall \alpha \in \Delta^+(\mathfrak{u}). \tag{2.32}$$

If α is a positive simple root for $\mathfrak{u}$, then $w\alpha$ can be of the form

$$w\alpha = \begin{cases} e_i - e_{i+1} & \text{for } i = \sum_{k=1}^{l}(p_k+q_k)\ ,\ 1 \le l \le r-1, \\ e_{n-m} - e_{n-m+1} & \text{if } m > 0, \\ 2e_n & \text{if } m = 0. \end{cases}$$

Equation 2.32 implies that

$$\lambda_1 \ge \lambda_2 \ge \cdots \ge \lambda_r \ge -\frac{1}{2}. \tag{2.33}$$

Then $w\gamma$ is weakly dominant with respect to $w\Delta^+(\mathfrak{g})$ (and of course γ is weakly dominant with respect to $\Delta^+(\mathfrak{g})$). Therefore, we can choose $\Delta^+(\gamma) = \Delta^+(\mathfrak{g})$. Conjugating $\gamma - \gamma_\omega^{\mathfrak{g}}$ in a similar way, we find

$$w(\gamma - \gamma_\omega^{\mathfrak{g}}) = \Big(\underbrace{\lambda_1+\frac{1}{2},\dots,\lambda_1+\frac{1}{2}}_{p_1+q_1},\underbrace{\lambda_2+\frac{1}{2},\dots,\lambda_2+\frac{1}{2}}_{p_2+q_2},\dots,\underbrace{\lambda_r+\frac{1}{2},\dots,\lambda_r+\frac{1}{2}}_{p_r+q_r},\underbrace{0,\dots,0}_{m}\Big).$$

Notice that the entries of $w(\gamma - \gamma_\omega^{\mathfrak{g}})$ are weakly decreasing and nonnegative (by 2.33). Hence $w(\gamma - \gamma_\omega^{\mathfrak{g}})$ is weakly dominant with respect to the roots in $w\Delta^+(\mathfrak{g})$:

$$\langle w(\gamma - \gamma_\omega^{\mathfrak{g}}), w\alpha\rangle \ge 0 \quad \forall \alpha \in \Delta^+(\mathfrak{g}).$$

Equivalently,

$$\langle \gamma - \gamma_\omega^{\mathfrak{g}}, \alpha\rangle \ge 0 \quad \forall \alpha \in \Delta^+(\mathfrak{g})$$

and γ is ω-regular. This concludes the proof of Proposition 3. □

2.3. Some Facts. The lowest K-types of the $A_{\mathfrak{q}}(\Omega)$ and Meta-$A_{\mathfrak{q}}(\lambda)$ representations will play a very important role in the rest of this paper.
Recall that the lowest K-types of a representation are those that are minimal with respect to the Vogan norm

$$\|\mu\| = (\mu + 2\rho_c, \mu + 2\rho_c)\,, \tag{2.34}$$

and that any irreducible representation has only finitely many lowest K-types.
It turns out that every $A_{\mathfrak{q}}(\Omega)$ and Meta-$A_{\mathfrak{q}}(\lambda)$ representation admits a unique lowest K-type, which is computed in the following proposition.

PROPOSITION 6. In the setting of Definitions 2.4 and 2.5, let $\rho(\mathfrak{u}\cap\mathfrak{p})$ be one half the sum of the noncompact roots of $\mathfrak{u}$. Then

(1) The $A_{\mathfrak{q}}(\Omega)$ representation $\mathcal{R}_{\mathfrak{q}}(\Omega)$ has a unique lowest K-type:

$$\mu = \mu(\mathfrak{q},\Omega) = \mu^L + 2\rho(\mathfrak{u}\cap\mathfrak{p}) \tag{2.35}$$

with μ^L the unique lowest $L\cap K$-type of Ω.

(2) The Meta-$A_{\mathfrak{q}}(\lambda)$ representation $\mathcal{R}_{\mathfrak{q}}(\mathbb{C}_\lambda\otimes J_\nu)$ has a unique lowest $L\cap K$-type:

$$\mu = \lambda\otimes 0 + 2\rho(\mathfrak{u}\cap\mathfrak{p}). \tag{2.36}$$

As usual, we have identified K-types and $L\cap K$-type with their highest weights.

PROOF. Both results follow from Proposition 5, because Ω and $\mathbb{C}_\lambda\otimes J_\nu$ are in the good range for $\mathfrak{q}$. □

To prove that certain representations are nonunitary, we will rely heavily on the following useful result.

PROPOSITION 7. (Parthasarathy's Dirac Operator Inequality [**8**], [**19**]) Let X be a unitary representation of G with infinitesimal character γ, and let μ be a K-type occurring in X. Choose a positive system $\Delta^+(\mathfrak{g},\mathfrak{t})\subseteq\Delta(\mathfrak{g},\mathfrak{t})$ of roots containing our fixed Δ_k^+, and let ρ_n,ρ_c be one half the sums of the noncompact and compact roots in $\Delta^+(\mathfrak{g},\mathfrak{t})$, respectively. Choose $w\in W_{\mathfrak{k}}$, the Weyl group of $\mathfrak{k}$, so that $w(\mu-\rho_n)$ is dominant with respect to Δ_k^+. Then

$$(w(\mu-\rho_n)+\rho_c, w(\mu-\rho_n)+\rho_c) \geq (\gamma,\gamma)\,. \tag{2.37}$$

We will often refer to the Parthasarathy's Dirac Operator Inequality as "PDOI".

If X is an irreducible admissible representation of $Mp(2n)$ and X^* its contragredient representation, then X and X^* have the same properties; in particular, X is unitary, ω-regular, an $A_{\mathfrak{q}}(\Omega)$, a Meta-$A_{\mathfrak{q}}(\lambda)$, a discrete series representation, finite dimensional, one-dimensional, etc. if and only if X^* is. We will sometimes use this symmetry to reduce the number of cases to be considered. Note that the K-types which occur in X^* are precisely those dual to the K-types occurring in X.

PROPOSITION 8. Let μ be an irreducible representation of $\widetilde{U}(n)$ with highest weight

$$\lambda = (a_1, a_{2,\dots,}a_n)\,. \tag{2.38}$$

Then the contragredient representation μ^* of μ has highest weight

$$\xi = (-a_n, -a_{n-1}, \cdots - a_2, -a_1)\,. \tag{2.39}$$

PROOF. Let μ be realized on the finite dimensional vector space V. Realize μ^* on the dual space V^*. The weights of μ^* are easily seen to be the opposite of the weights of μ: if $\{v_{\lambda_1}, v_{\lambda_2},\dots,v_{\lambda_r}\}$ is a basis of V consisting of weight vectors corresponding to the weights $\lambda_1,\dots,\lambda_r$, then the dual basis $\{v^*_{\lambda_1}, v^*_{\lambda_2},\dots,v^*_{\lambda_r}\}$ of V^* is a set of weight vectors corresponding to the weights $-\lambda_1,\dots,-\lambda_r$.

Define $-\xi = (a_n, a_{n-1},\dots,a_1)$, with $\lambda = (a_1, a_{2,\dots,}a_n)$ the highest weight for μ.

$-\xi$ is an extremal weight of μ (because is Weyl group conjugate to λ), and is the lowest weight of μ (because it is antidominant with respect to our fixed set of positive roots). Then $\zeta - (-\xi)$ is a a sum of positive roots, for every weight ζ of μ. Equivalently, $\xi - (-\zeta)$ is a a sum of positive roots, for every weight $(-\zeta)$ of μ^*, hence ξ is the highest weight of μ^*. □

2.4. Langlands Classification and Lowest K-Types. Our proof of Theorem 2.8 proceeds by K-types: for each K-type μ and each ω-regular irreducible unitary representation π with lowest K-type μ, we show that π must be either an $A_{\mathfrak{q}}(\Omega)$ or a Meta-$A_{\mathfrak{q}}(\lambda)$ representation. Therefore, it is important to know which representations contain a given μ as a lowest K-type, and what are the possible infinitesimal characters for such representations.
Because all ω-regular infinitesimal characters are in particular real, we will assume from now on that all infinitesimal characters have this property.

To determine the set of representations with a given infinitesimal character, we use the Langlands Classification, which is a construction equivalent to Vogan's construction from Proposition 1, but uses real parabolic induction instead of cohomological parabolic induction. According to the Langlands Classification (c.f. [**5**], [**17**]), every irreducible admissible representation of G occurs as an irreducible quotient $\overline{X}(\sigma, \nu)$ of an induced representation

$$I_P(\sigma, \nu) = \mathrm{Ind}_P^G(\sigma \otimes \nu \otimes 1), \tag{2.40}$$

where $P = MAN$ is a cuspidal parabolic subgroup of G, σ a discrete series representation of M, and ν a character of A. (We are abusing notation and using ν to denote both the character and its differential.) If the infinitesimal character of the representation is regular, as it always is in our setting, then $I_P(\sigma, \nu)$ has a unique irreducible quotient. Inducing data give rise to equivalent irreducible representations if and only if they are conjugate by G.

We now give a more specific description of the data for irreducible representations of $Mp(4)$ with real regular infinitesimal character (see also [**1**], [**9**]); for $Mp(2)$, the situation is similar, yet much simpler.
Irreducible representations of $Mp(4)$ are in one-one correspondence with triples

$$(MA,\ \sigma,\ \nu) \tag{2.41}$$

as follows. There are four conjugacy classes of cuspidal parabolic subgroups, given by their Levi factors

$$MA = \begin{cases} Mp(4) & \text{with } \ M = Mp(4),\ A = \{1\} \\ Mp(2) \times \widetilde{GL}(1,\mathbb{R}) & \text{with } \ M = Mp(2) \times \mathbb{Z}/4\mathbb{Z},\ A = \mathbb{R} \\ \widetilde{GL}(2,\mathbb{R}) & \text{with } \ M = Mp(2)^{\pm},\ A = \mathbb{R} \\ \widetilde{GL}(1,\mathbb{R})^2 & \text{with } \ M = (\mathbb{Z}/4\mathbb{Z})^2,\ A = \mathbb{R}^2 \end{cases} \tag{2.42}$$

(in the second and fourth case, MA and M are actually quotients by a subgroup of order 2 of this product). The group $\widetilde{GL}(2,\mathbb{R})$ above is the split double cover of $GL(2,\mathbb{R})$. The discrete series σ may be given by its Harish-Chandra parameter and by a character of $\mathbb{Z}/4\mathbb{Z}$ or $(\mathbb{Z}/4\mathbb{Z})^2$. The parameter ν can be conjugated into a positive number or a pair of positive numbers (recall that we are only considering representations with real regular infinitesimal character, so ν is real and, if A is nontrivial, ν is nonzero).

Recall that to every K-type μ we can assign a Vogan parameter $\lambda_a = \lambda_a(\mu) \in \mathfrak{t}^*$ as follows (c.f. [**16**]): choose a representative of ρ such that $\mu + 2\rho_c$ is weakly dominant with respect to ρ. Then

$$\lambda_a = p(\mu + 2\rho_c - \rho), \tag{2.43}$$

where p denotes the projection onto the positive Weyl chamber determined by ρ. The Vogan parameter λ_a then gives the Harish-Chandra parameter of σ, for any representation $\overline{X}(\sigma, \nu)$ with lowest K-type μ. This determines the (conjugacy class of the) Levi subgroup MA as well. Write

$$\lambda_a = (a, b) \tag{2.44}$$

with $a \geq b$.

(1) If $|a| \neq |b|$ and both are nonzero, then $MA = Mp(4)$, and λ_a is the Harish-Chandra parameter of a discrete series of G.

(2) If $a > b = 0$ then $MA = Mp(2) \times \widetilde{GL}(1, \mathbb{R})$. In this case, $A = \mathbb{R}$ so ν is just a positive number. The infinitesimal character of the corresponding representation is

$$\gamma = (a, \nu). \tag{2.45}$$

The character of $\mathbb{Z}/4\mathbb{Z}$ is also uniquely determined by μ. An analogous statement holds if $a = 0 > b$; then we have $\gamma = (\nu, b)$.

(3) If $a = -b \neq 0$ then $MA = \widetilde{GL}(2, \mathbb{R})$. Also in this case, $A = \mathbb{R}$ and ν is a positive number. The infinitesimal character of the corresponding representation is

$$\gamma = (a + \nu, -a + \nu). \tag{2.46}$$

(4) If $a = b = 0$, then we say that μ is a fine K-type. In this case, the representation is a principal series, and has infinitesimal character

$$\gamma = (\nu_1, \nu_2). \tag{2.47}$$

The case $a = b \neq 0$ does not occur; the parameter λ_a must be such that its centralizer L_a in G is a quasisplit Levi subgroup [**16**].

3. The Group Mp(2)

Let $G = Mp(2, \mathbb{R})$ be the connected double cover of

$$Sp(2, \mathbb{R}) = \left\{ g \in GL(2, \mathbb{R}) : g^t \begin{pmatrix} 0 & 1 \\ -1 & 0 \end{pmatrix} g = \begin{pmatrix} 0 & 1 \\ -1 & 0 \end{pmatrix} \right\}. \tag{3.1}$$

Note that $Sp(2, \mathbb{R})$ equals $SL(2, \mathbb{R})$. The Lie algebra of G is

$$\mathfrak{g}_0 = \left\{ \begin{pmatrix} a & b \\ c & -a \end{pmatrix} : a, b, c \in \mathbb{R} \right\}, \tag{3.2}$$

and the maximal compact Cartan subalgebra of $\mathfrak{g}_0$ is

$$\mathfrak{t}_0 = \mathfrak{k}_0 = \left\{ \begin{pmatrix} 0 & t \\ -t & 0 \end{pmatrix} : t \in \mathbb{R} \right\}. \tag{3.3}$$

The maximal compact subgroup of $Sp(2, \mathbb{R})$ is $SO(2) \simeq U(1)$, hence the maximal compact subgroup K of $Mp(2, \mathbb{R})$ is isomorphic to $\widetilde{U}(1)$. We identify $\widehat{K}$ with $\frac{1}{2}\mathbb{Z}$, as follows: write

$$K \simeq \left\{ (g, z) \in U(1) \times \mathbb{C}^\times : z^2 = g \right\}. \tag{3.4}$$

Then the character of K corresponding to the half integer a is given by

$$(g,z) \mapsto z^{2a}. \tag{3.5}$$

This character is genuine if and only if $a \in \mathbb{Z} + \frac{1}{2}$.

We are interested in ω-regular unitary representations of G. It turns out that they are all obtained by either complementary series or cohomological parabolic induction from a Levi subgroup L of a theta stable parabolic subalgebra $\mathfrak{q} = \mathfrak{l} + \mathfrak{u}$ of $\mathfrak{g}$. This subalgebra $\mathfrak{q}$ is related to the subalgebra $\mathfrak{q}_a$ defined in Proposition 1.

PROPOSITION 9. If $G = Mp\,(2,\mathbb{R})$ and X is an irreducible unitary representation of G with ω-regular infinitesimal character, then either

(1) $X \simeq A_{\mathfrak{q}}(\Omega)$ for some θ-stable parabolic subalgebra $\mathfrak{q}$ and some representation Ω of L, as in Definition 2.4, or
(2) X is isomorphic to a Meta-$A_{\mathfrak{q}}(\lambda)$ representation, as in Definition 2.5.

PROOF. Let X be an irreducible unitary representation of G with ω-regular infinitesimal character, and let $\mu = (a) \in \frac{1}{2}\mathbb{Z}$ be a lowest K-type for X. We prove that X is either an $A_q(\Omega)$ or a Meta-$A_q(\lambda)$ representation. Recall that, for $Mp(2)$, the $A_q(\Omega)$ representations are the oscillator representations and the genuine discrete series. The Meta-$A_q(\lambda)$ representations are the nongenuine discrete series and the complementary series J_ν, with $\frac{1}{2} \leq \nu \leq 1$ (if $\nu = 1$ then J_ν is trivial representation).

First assume that $a \in \frac{1}{2}\mathbb{Z} \setminus \{0, \pm\frac{1}{2}, \pm 1\}$. Vogan's classification of irreducible admissible representations ([**15**]) implies that if X has lowest K-type $\mu = (a)$ in $\frac{1}{2}\mathbb{Z} \setminus \{0, \pm\frac{1}{2}, \pm 1\}$, then X is a discrete series representation with Harish-Chandra parameter $\lambda = a - sgn\,(a) \neq 0$. Hence X is an $A_q(\Omega)$ representation if genuine, and a Meta-$A_q(\lambda)$ representation if nongenuine. We notice that in this case

$$X = A_{\mathfrak{q}}(\lambda) \tag{3.6}$$

with $\mathfrak{q}$ the Borel subalgebra determined by λ. Because $|\lambda| \geq \frac{1}{2}$, X is ω-regular (see Example 2.3).

We are left with the cases $a \in \{0, \pm\frac{1}{2}, \pm 1\}$. First assume $a = \pm 1$, and choose $\rho_n = \pm 1$. Since $\rho_c = 0$, Vogan's classification of admissible representations gives that X has infinitesimal character $\gamma = (\lambda_a, \nu) = (0, \nu)$. In this case, for w trivial, we have

$$\langle w(\mu - \rho_n) + \rho_c,\, w(\mu - \rho_n) + \rho_c \rangle = 0 \tag{3.7}$$

so the Parthasarathy's Dirac operator inequality ("PDOI", cf. Proposition 7) yields that if $\nu \neq 0$, then X is nonunitary. We conclude that there are no irreducible unitary ω-regular representations of G with lowest K-type $\mu = \pm 1$.

Next, assume $a = \pm\frac{1}{2}$. Note that $\mu = \pm\frac{1}{2}$ is the lowest K-type of an even oscillator representation ω, and that $\omega = A_{\mathfrak{q}}(\Omega)$ with $\mathfrak{q} = \mathfrak{g}$ and $\Omega = \omega$. We will show that the oscillator representations are the only irreducible unitary ω-regular representations X of G containing $\mu = \pm\frac{1}{2}$ as their lowest K-type.
Choose $\mu = \pm\frac{1}{2}$, and $\rho_n = \pm 1$ with the same sign as μ. For w trivial, we get

$$\langle w(\mu - \rho_n) + \rho_c,\, w(\mu - \rho_n) + \rho_c \rangle = \frac{1}{4}. \tag{3.8}$$

So PDOI implies that X is nonunitary if its infinitesimal character γ satisfies $\langle \gamma, \gamma \rangle > \frac{1}{4}$. On the other hand, X is not ω-regular if $\langle \gamma, \gamma \rangle < \frac{1}{4}$. Hence any representation X of G (with lowest K-type $\pm\frac{1}{2}$) which is both ω-regular and unitary must satisfy $\langle \gamma, \gamma \rangle = \frac{1}{4}$.

Because $\mu = \left(\pm\frac{1}{2}\right)$ is fine, and X contains μ as a lowest K-type, X must be induced from a representation $\delta \otimes \nu$ of $P = MAN$, where

$$M \simeq \widetilde{\mathbb{Z}_2} \simeq \mathbb{Z}_4. \tag{3.9}$$

The infinitesimal character γ of X is given by $(0, \nu)$, and the condition $\langle\gamma, \gamma\rangle = \frac{1}{4}$ implies $\nu = \pm\frac{1}{2}$. The two choices are conjugate by the Weyl group, hence give equivalent representations; we assume $\nu = \frac{1}{2}$. Next, we prove that the choice of δ is also uniquely determined by μ. This is an easy application of Frobenius reciprocity: if μ is contained in $X = \mathrm{Ind}(\delta \otimes \nu)$, then δ is contained in the restriction of μ to M. With our identification (3.4), we can write

$$M = \{(1, \pm 1), (-1, \pm i)\} \tag{3.10}$$

and the restriction of a K-type $\mu = (b)$ to M is the character

$$\widetilde{U}(1) \to \mathbb{C}^{\times},\ (g, z) \mapsto z^{2b}.$$

The K-types $\mu = \frac{1}{2}$ and $\mu = -\frac{1}{2}$ restrict to the characters $(g, z) \mapsto z$ and $(g, z) \mapsto z^{-1}$ respectively. Then δ must be the identity M-type $(x \mapsto x)$ if $\mu = \frac{1}{2}$, and the inverse M-type $(x \mapsto x^{-1})$ if $\mu = -\frac{1}{2}$. Note that, in both cases, $X = \mathrm{Ind}(\delta \otimes \frac{1}{2})$ is an oscillator representation.

Lastly, we assume $a = 0$. If X is an irreducible unitary ω-regular representation containing the trivial K-type, then $X = J_\nu$ for some value of ν (these are the only spherical representations of $Mp(2)$). Note that PDOI implies that $|\nu| \leq 1$ and the ω-regular condition requires that $|\nu| \geq \frac{1}{2}$, hence $\frac{1}{2} \leq \nu \leq 1$. So X is a complementary series (and a Meta-$A_q(\lambda)$ representation). □

4. The Group Mp(4)

4.1. The Structure of Mp(4). We realize the Lie algebra $\mathfrak{g}_0 = \mathfrak{sp}(4, \mathbb{R})$ of $Mp(4)$ and the Lie algebra $\mathfrak{k}_0$ of its maximal compact subgroup as

$$\mathfrak{g}_0 = \left\{ \left(\begin{array}{cc|cc} a_{11} & a_{12} & b_{11} & b_{12} \\ a_{21} & a_{22} & b_{12} & b_{22} \\ \hline c_{11} & c_{12} & -a_{11} & -a_{21} \\ c_{12} & c_{22} & -a_{12} & -a_{22} \end{array} \right) : a_{i,j}, b_{i,j}, c_{i,j} \in \mathbb{R} \right\} \tag{4.1}$$

and

$$\mathfrak{k}_0 = \left\{ Y = \left(\begin{array}{cc|cc} 0 & a & x & z \\ -a & 0 & z & y \\ \hline -x & -z & 0 & a \\ -z & -y & -a & 0 \end{array} \right) : a, x, y, z \in \mathbb{R} \right\}. \tag{4.2}$$

The maximal compact subgroup $K \simeq \widetilde{U}(2)$ of $G = Mp(4)$ is isomorphic to the subgroup

$$U = \{(g, z) \in U(2) \times U(1)\colon\ \det(g) = z^2\}$$

of $U(2) \times U(1)$. Identifying $\mathfrak{k}_0$ with the Lie algebra of U gives a map

$$\iota\colon \mathfrak{k}_0 \to \mathfrak{u}(2) \oplus \mathfrak{u}(1),\ Y \mapsto \iota(Y) = \left[\begin{pmatrix} xi & a + zi \\ -a + zi & yi \end{pmatrix}, \frac{x+y}{2} i \right], \tag{4.3}$$

where Y is the element of $\mathfrak{k}_0$ as in (4.2). We denote the exponentiated map $K \to U$ by ι as well. Let $\mathfrak{a}_0$ be the diagonal CSA of $\mathfrak{g}_0$, $A = \exp(\mathfrak{a}_0)$, and $M = Cent_K(A)$. Here exp denotes the exponential map in $Mp(4)$. Also let

$$a = \exp\left(\begin{array}{cc|cc} 0 & 0 & \pi & 0 \\ 0 & 0 & 0 & 0 \\ \hline -\pi & 0 & 0 & 0 \\ 0 & 0 & 0 & 0 \end{array}\right) \qquad b = \exp\left(\begin{array}{cc|cc} 0 & 0 & 0 & 0 \\ 0 & 0 & 0 & \pi \\ \hline 0 & 0 & 0 & 0 \\ 0 & -\pi & 0 & \end{array}\right), \tag{4.4}$$

$$x = \exp\left(\begin{array}{cc|cc} 0 & 0 & \pi & 0 \\ 0 & 0 & 0 & \pi \\ \hline -\pi & 0 & 0 & 0 \\ 0 & -\pi & 0 & 0 \end{array}\right) \qquad y = \exp\left(\begin{array}{cc|cc} 0 & 0 & \pi & 0 \\ 0 & 0 & 0 & -\pi \\ \hline -\pi & 0 & 0 & 0 \\ 0 & \pi & 0 & 0 \end{array}\right). \tag{4.5}$$

Then

$$\iota(a) = \exp\left[\begin{pmatrix} i\pi & 0 \\ 0 & 0 \end{pmatrix}, \frac{\pi}{2}i\right] = \left[\begin{pmatrix} -1 & 0 \\ 0 & 1 \end{pmatrix}, i\right] \tag{4.6}$$

$$\iota(b) = \exp\left[\begin{pmatrix} 0 & 0 \\ 0 & i\pi \end{pmatrix}, \frac{\pi}{2}i\right] = \left[\begin{pmatrix} 1 & 0 \\ 0 & -1 \end{pmatrix}, i\right] \tag{4.7}$$

$$\iota(x) = \exp\left[\begin{pmatrix} i\pi & 0 \\ 0 & i\pi \end{pmatrix}, \pi i\right] = \left[\begin{pmatrix} -1 & 0 \\ 0 & -1 \end{pmatrix}, -1\right] \quad \text{and} \tag{4.8}$$

$$\iota(y) = \exp\left[\begin{pmatrix} \pi i & 0 \\ 0 & -\pi i \end{pmatrix}, 0\right] = \left[\begin{pmatrix} -1 & 0 \\ 0 & -1 \end{pmatrix}, 1\right]. \tag{4.9}$$

We notice that $pr(a) = diag(-1, 1, -1, 1) = pr(a^{-1})$, $x = ab$ and $y = ab^{-1}$. Set

$$z := a^2 = \exp\left(\begin{array}{cc|cc} 0 & 0 & 2\pi & 0 \\ 0 & 0 & 0 & 0 \\ \hline -2\pi & 0 & 0 & 0 \\ 0 & 0 & 0 & 0 \end{array}\right) = \exp\left(\begin{array}{cc|cc} 0 & 0 & 0 & 0 \\ 0 & 0 & 0 & 2\pi \\ \hline 0 & 0 & 0 & 0 \\ 0 & -2\pi & 0 & 0 \end{array}\right) = b^2. \tag{4.10}$$

Then $Z(Mp(4)) = \{e, z, x, y\} = \langle x, y\rangle \simeq \mathbb{Z}_2 \times \mathbb{Z}_2$, with e the identity, and

$$M = \{e, z, x, y, a, a^{-1}, b, b^{-1}\} = \langle a, b\rangle \simeq \mathbb{Z}_4 \times \mathbb{Z}_2. \tag{4.11}$$

Let

$$\mathfrak{t}_0 = \left\{\left(\begin{array}{cc|cc} 0 & 0 & \theta & 0 \\ 0 & 0 & 0 & \varphi \\ \hline -\theta & 0 & 0 & 0 \\ 0 & -\varphi & 0 & 0 \end{array}\right) : \theta, \varphi \in \mathbb{R}\right\} \subseteq \mathfrak{k}_0, \tag{4.12}$$

a fundamental Cartan subalgebra. Then

$$\iota(\mathfrak{t}_0) = \left\{\left[\begin{pmatrix} \theta i & 0 \\ 0 & \varphi i \end{pmatrix}, \frac{\theta + \varphi}{2} i\right] : \theta, \varphi \in \mathbb{R}\right\} \tag{4.13}$$

and the corresponding Cartan subgroup T is given by

$$\iota(T) = \left\{t_{\theta,\varphi} = \left[\begin{pmatrix} e^{i\theta} & 0 \\ 0 & e^{i\varphi} \end{pmatrix}, e^{\frac{\theta+\varphi}{2} i}\right] : \theta, \varphi \in \mathbb{R}\right\}. \tag{4.14}$$

A weight $\mu = \left(\frac{k}{2}, \frac{l}{2}\right)$ corresponds to the character of T given by

$$t_{\theta,\varphi} \mapsto e^{\frac{k\theta + l\varphi}{2} i}. \tag{4.15}$$

We identify K-types with their highest weights. The K-type

$$\mu = \left(\frac{k}{2}, \frac{l}{2}\right) \tag{4.16}$$

is genuine if and only if both k and l are odd, and nongenuine if they are both even. In each case, the K-type has dimension $\frac{k-l}{2}+1$, the other weights being

$$\left(\frac{k}{2} - j, \frac{l}{2} + j\right), \text{ for } 1 \leq j \leq \frac{k-l}{2}. \tag{4.17}$$

We will check the unitarity of (ω-regular) representations of $Mp(4)$ by partitioning the set of $\widetilde{U}(2)$-types μ in a suitable way, and considering for each family the set of ω-regular representations which have such a μ as a lowest K-type. The symmetry considerations at the end of Section 2.3 and Proposition 8 reduce the $\widetilde{U}(2)$-types we need to consider to those of the form

$$\mu = \left(\frac{k}{2}, \frac{l}{2}\right) \text{ with } k \geq |l|. \tag{4.18}$$

4.2. The Genuine Case: K-Types. We partition the genuine K-types into those that are lowest K-types of $A_{\mathfrak{q}}(\Omega)$ representations, and those that are not. In order to construct an $A_{\mathfrak{q}}(\Omega)$ module, we must start with a theta stable subalgebra $\mathfrak{q} = \mathfrak{l} + \mathfrak{u}$. Any such algebra is of the form

$$\mathfrak{q} = \mathfrak{q}(\xi) = \mathfrak{l}(\xi) + \mathfrak{u}(\xi),$$

where $\xi \in i\mathfrak{t}_0^*$,

$$\mathfrak{l}(\xi) = \mathfrak{t} + \sum_{\langle \alpha, \xi \rangle = 0} \mathfrak{g}_\alpha,$$

$$\mathfrak{u}(\xi) = \sum_{\langle \alpha, \xi \rangle > 0} \mathfrak{g}_\alpha.$$

We get 10 theta stable parabolic algebras $\mathfrak{q}_i = \mathfrak{q}(\xi_i)$, with

$$\begin{array}{ll} \xi_1 = (0,0) & \xi_6 = (0,-1) \\ \xi_2 = (2,1) & \xi_7 = (-1,-2) \\ \xi_3 = (2,-1) & \xi_8 = (1,-2) \\ \xi_4 = (1,-1) & \xi_9 = (1,1) \\ \xi_5 = (1,0) & \xi_{10} = (-1,-1). \end{array}$$

The corresponding Levi factors L_i are

$$\begin{array}{ll} L_1 = Mp(4) & L_6 = \widetilde{U}(0,1) \times Mp(2) \\ L_2 = \widetilde{U}(1,0) \times \widetilde{U}(1,0) & L_7 = \widetilde{U}(0,1) \times \widetilde{U}(0,1) \\ L_3 = \widetilde{U}(1,0) \times \widetilde{U}(0,1) & L_8 = \widetilde{U}(0,1) \times \widetilde{U}(1,0) \\ L_4 = \widetilde{U}(1,1) & L_9 = \widetilde{U}(2,0) \\ L_5 = \widetilde{U}(1,0) \times Mp(2) & L_{10} = \widetilde{U}(0,2). \end{array}$$

Some of these Levi subgroups are, of course, pairwise identical; however, the corresponding nilpotent parts of the parabolic subalgebras are different. We express these differences in our notation for the L_i as above. We notice that

- any $A_{\mathfrak{q}}(\Omega)$ representation with $L = L_7$ is dual to one with $L = L_2$. Similarly for the pairs $\{L_6, L_5\}$, $\{L_8, L_3\}$, and $\{L_{10}, L_9\}$;

- any $A_{\mathfrak{q}}(\Omega)$ representation with $L = L_9$ is a discrete series which may also be constructed with $L = L_2$ (see [**6**]).

Hence we may restrict our attention to the cases

$$L_1 = Mp(4), \tag{4.19}$$

$$L_2 = \widetilde{U}(1,0) \times \widetilde{U}(1,0), \tag{4.20}$$

$$L_3 = \widetilde{U}(1,0) \times \widetilde{U}(0,1), \tag{4.21}$$

$$L_4 = \widetilde{U}(1,1) \text{ and} \tag{4.22}$$

$$L_5 = \widetilde{U}(1,0) \times Mp(2). \tag{4.23}$$

With $L_1 = Mp(4)$, the $A_{\mathfrak{q}}(\Omega)$ modules we obtain are the four oscillator representations with lowest K-types

$$\Lambda_1 = \left\{ \left(\frac{3}{2}, \frac{1}{2}\right), \left(\frac{1}{2}, \frac{1}{2}\right), \left(-\frac{1}{2}, -\frac{1}{2}\right), \left(-\frac{1}{2}, -\frac{3}{2}\right) \right\}. \tag{4.24}$$

Now consider $L_2 = \widetilde{U}(1,0) \times \widetilde{U}(1,0)$. Then

$$\rho(\mathfrak{u}) = (2,1) \tag{4.25}$$

$$\rho(\mathfrak{l}) = (0,0) \tag{4.26}$$

$$\rho(\mathfrak{u} \cap \mathfrak{p}) = \left(\frac{3}{2}, \frac{3}{2}\right). \tag{4.27}$$

Set

$$\lambda = (\lambda_1, \lambda_2) \tag{4.28}$$

with $\lambda_i \in \mathbb{Z} + \frac{1}{2}$. For λ to be in the good range for $\mathfrak{q}$, the parameter

$$\lambda + \rho(\mathfrak{l}) + \rho(\mathfrak{u}) = (\lambda_1 + 2, \lambda_2 + 1) \tag{4.29}$$

must be strictly dominant for the roots $\{2e_1, 2e_2, e_1 \pm e_2\}$ of $\mathfrak{u}$. This says that

$$\lambda_1 + 2 > \lambda_2 + 1 > 0 \quad \Leftrightarrow \quad \lambda_1 \geq \lambda_2 \geq -\frac{1}{2}. \tag{4.30}$$

We obtain lowest K-types of the form

$$\mu = \lambda + 2\rho(\mathfrak{u} \cap \mathfrak{p}) = (\lambda_1 + 3, \lambda_2 + 3)\,, \tag{4.31}$$

which belong to the set

$$\Lambda_2 = \left\{ (r,s) \colon r \geq s \geq \frac{5}{2} \right\}. \tag{4.32}$$

For $L_3 = \widetilde{U}(1,0) \times \widetilde{U}(0,1)$, we have essentially two choices for $\mathfrak{u}$, corresponding to

$$\rho(\mathfrak{u}) = (2,-1) \quad \text{and} \quad \rho(\mathfrak{u}) = (1,-2). \tag{4.33}$$

The collection of $A_{\mathfrak{q}}(\Omega)$ modules obtained with the second choice are easily seen to be the contragredient modules of those obtained with the first choice, so we may assume that

$$\rho(\mathfrak{u}) = (2,-1) \tag{4.34}$$

$$\rho(\mathfrak{l}) = (0,0) \tag{4.35}$$

$$\rho(\mathfrak{u}\cap\mathfrak{p}) = \left(\frac{3}{2},-\frac{1}{2}\right). \tag{4.36}$$

This time $\lambda = (\lambda_1, \lambda_2)$ is in the good range if the parameter

$$\lambda + \rho(\mathfrak{l}) + \rho(\mathfrak{u}) = (\lambda_1 + 2, \lambda_2 - 1) \tag{4.37}$$

is strictly dominant for the roots $\{2e_1, -2e_2, e_1 \pm e_2\}$, i. e.,

$$\lambda_1 + 2 > -(\lambda_2 - 1) > 0 \quad \Leftrightarrow \quad \lambda_1 \geq -\lambda_2 \geq -\frac{1}{2}. \tag{4.38}$$

We get lowest K-types of the form

$$\mu = (\lambda_1 + 3, \lambda_2 - 1), \tag{4.39}$$

which give rise to the collection

$$\Lambda_3 = \left\{(r,s) : s \leq -\frac{1}{2}, r \geq -s+2\right\}. \tag{4.40}$$

For $L_4 = \widetilde{U}(1,1)$, we have

$$\rho(\mathfrak{u}) = \left(\frac{3}{2},-\frac{3}{2}\right) \tag{4.41}$$

$$\rho(\mathfrak{l}) = \left(\frac{1}{2},\frac{1}{2}\right) \tag{4.42}$$

$$\rho(\mathfrak{u}\cap\mathfrak{p}) = (1,-1). \tag{4.43}$$

Set

$$\lambda = (\lambda_1, -\lambda_1) \tag{4.44}$$

with $\lambda_1 \in \mathbb{Z} + \frac{1}{2}$. Then λ is in the good range if

$$\lambda + \rho(\mathfrak{l}) + \rho(\mathfrak{u}) = (\lambda_1 + 2, -\lambda_1 - 1) \tag{4.45}$$

is strictly dominant for the roots $\{2e_1, -2e_2, e_1 - e_2\}$ of $\mathfrak{u}$. Given that λ_1 is half-integral, this condition is equivalent to

$$\lambda_1 \geq -\frac{1}{2}. \tag{4.46}$$

We obtain lowest K-types of the form

$$\mu = \lambda + 2\rho(\mathfrak{u}\cap\mathfrak{p}) = (\lambda_1 + 2, -\lambda_1 - 2), \tag{4.47}$$

which belong to the set

$$\Lambda_4 = \left\{(r,-r) : r \geq \frac{3}{2}\right\}. \tag{4.48}$$

Finally, let $L_5 = \widetilde{U}(1,0) \times Mp(2)$. In this case we have

$$\rho(\mathfrak{u}) = (2,0) \tag{4.49}$$

$$\rho(\mathfrak{l}) = (0,1) \tag{4.50}$$

$$\rho(\mathfrak{u}\cap\mathfrak{p}) = \left(\frac{3}{2},\frac{1}{2}\right). \tag{4.51}$$

We consider representations of L_5 of the form $\Omega = \mathbb{C}_\lambda \otimes \omega$, with ω an oscillator representation of $Mp(2)$. If ω is an *odd* oscillator representations, then ω is a discrete series of $Mp(2)$; the corresponding $A_\mathfrak{q}(\Omega)$ representation is a discrete series of $Mp(4)$ and has already been considered. Hence we may assume that ω is an *even* oscillator representation of $Mp(2)$.
The infinitesimal character of $\Omega = \mathbb{C}_\lambda \otimes \omega$ is $\left(\lambda, \frac{1}{2}\right)$; this is in the good range if

$$\left(\lambda, \frac{1}{2}\right) + (2, 0) = \left(\lambda + 2, \frac{1}{2}\right) \tag{4.52}$$

is strictly dominant with respect to $\Delta(\mathfrak{u}) = \{2e_1, e_1 \pm e_2\}$, i. e.,

$$\lambda \geq -\frac{1}{2}. \tag{4.53}$$

The lowest $L \cap K$-type of Ω is then either

$$\left(\lambda, \frac{1}{2}\right) \text{ or } \left(\lambda, -\frac{1}{2}\right) \tag{4.54}$$

depending on whether ω is even holomorphic or even antiholomorphic. Adding $2\rho(\mathfrak{u} \cap \mathfrak{p})$ we get the following set of lowest K-types

$$\Lambda_5 = \left\{\left(r, \frac{3}{2}\right) : r \geq \frac{5}{2}\right\} \cup \left\{\left(r, \frac{1}{2}\right) : r \geq \frac{5}{2}\right\}. \tag{4.55}$$

REMARK 4.1. The sets Λ_i for $1 \leq i \leq 5$ list all the genuine K-types (r, s) which occur as lowest K-types of $A_\mathfrak{q}(\Omega)$ modules, and satisfy $r \geq |s|$.

This leaves us with the following set Σ of genuine K-types which are NOT lowest K-types of $A_\mathfrak{q}(\Omega)$ representations:

$$\Sigma = \left\{\left(\frac{1}{2}, -\frac{1}{2}\right), \left(\frac{3}{2}, \frac{3}{2}\right)\right\} \cup \left\{(r, -r+1) : r \geq \frac{3}{2}\right\}. \tag{4.56}$$

4.3. The Set Σ of Non-$A_\mathfrak{q}(\Omega)$ Lowest K-Types. We show that any ω-regular representation with a lowest K-type in the set Σ must be nonunitary. In most cases this is done using PDOI.

First consider

$$\mu = \left(\frac{3}{2}, \frac{3}{2}\right). \tag{4.57}$$

Choose

$$\rho_n = \left(\frac{3}{2}, \frac{3}{2}\right) = \mu \tag{4.58}$$

and w trivial. Then

$$\langle w(\mu - \rho_n) + \rho_c, w(\mu - \rho_n) + \rho_c \rangle = \langle \rho_c, \rho_c \rangle = \frac{1}{2} \tag{4.59}$$

because $\rho_c = \left(\frac{1}{2}, -\frac{1}{2}\right)$. Notice that any ω-regular infinitesimal character γ satisfies

$$\langle \gamma, \gamma \rangle \geq \langle \gamma_\omega, \gamma_\omega \rangle = \left(\frac{3}{2}\right)^2 + \left(\frac{1}{2}\right)^2 = \frac{5}{2} > \frac{1}{2}, \tag{4.60}$$

hence the Parthasarathy's Dirac operator inequality (2.37) fails. This proves that any ω-regular representation with lowest K-type $\mu = \left(\frac{3}{2}, \frac{3}{2}\right)$ is nonunitary.

Now let

$$\mu = (r, -r+1) \tag{4.61}$$

with $r \geq \frac{3}{2}$. Choose

$$\rho_n = \left(\frac{3}{2}, -\frac{1}{2}\right) \tag{4.62}$$

and w trivial. Then

$$w(\mu - \rho_n) + \rho_c = (r-1, -r+1)\,. \tag{4.63}$$

and

$$\langle w(\mu - \rho_n) + \rho_c, w(\mu - \rho_n) + \rho_c \rangle = 2(r-1)^2. \tag{4.64}$$

A representation with this lowest K-type has Vogan parameter

$$\lambda_a = p(\mu + 2\rho_c - \rho). \tag{4.65}$$

Since

$$\rho = (2, -1), \tag{4.66}$$

we get

$$\lambda_a = (r-1, -r+1). \tag{4.67}$$

Hence the corresponding standard module is induced from a parabolic subgroup $P = MAN$ with $MA \simeq GL(2, \mathbb{R})$, and the infinitesimal character is of the form

$$\gamma = (r - 1 + \nu, -r + 1 + \nu) \tag{4.68}$$

for some number ν (see section 2.4).
In order for γ to be ω-regular, ν must be real with $|\nu| \geq \frac{1}{2}$. Recall from Section 2.4 that we may conjugate ν to be positive. Then we have $\nu \geq 1/2$, and

$$\langle \gamma, \gamma \rangle = (r - 1 + \nu)^2 + (r - 1 - \nu)^2 > 2(r-1)^2 \tag{4.69}$$

Because the PDOI fails, such a representation is nonunitary.

Finally consider the K-type

$$\mu = \left(\frac{1}{2}, -\frac{1}{2}\right). \tag{4.70}$$

If we choose

$$\rho_n = \left(\frac{3}{2}, -\frac{1}{2}\right) \tag{4.71}$$

and $w = -1$ (the long Weyl group element), then

$$w(\mu - \rho_n) + \rho_c = \left(\frac{3}{2}, \frac{1}{2}\right), \tag{4.72}$$

and

$$\langle w(\mu - \rho_n) + \rho_c, w(\mu - \rho_n) + \rho_c \rangle = \frac{5}{2} = \langle \gamma_\omega, \gamma_\omega \rangle. \tag{4.73}$$

PDOI implies that any unitary ω-regular representation containing this K-type must have infinitesimal character γ_ω. It is easy to check that other choices for ρ_n do not give any better estimate. So it remains to show that the irreducible representations with lowest K-type $\mu = \left(\frac{1}{2}, -\frac{1}{2}\right)$ and infinitesimal character γ_ω (these are two principal series representations which are dual to each other) are nonunitary. We do this in Section 5, by explicitly computing the signature of a Hermitian form.

4.4. Uniqueness of Representations with $A_{\mathfrak{q}}(\Omega)$ Lowest K-types. In this section we prove that there are no ω-regular unitary representations of G which have an $A_{\mathfrak{q}}(\Omega)$ lowest K-type, but are not $A_{\mathfrak{q}}(\Omega)$ representations. We do this case by case, considering in turn the K-types listed in the sets Λ_i in Section 4.2. For each K-type μ, we show that there is only one ω-regular irreducible representation with lowest K-type μ and such that μ satisfies PDOI. We rely heavily on the Langlands classification and Vogan's lowest K-type ideas as outlined in Section 2.4.

4.4.1. *The Set* Λ_1. Suppose $\mu \in \Lambda_1$. We only have to consider the cases $\mu = \left(\frac{1}{2}, \frac{1}{2}\right)$ and $\mu = \left(\frac{3}{2}, \frac{1}{2}\right)$ (the other K-types in Λ_1 are dual to these, by Proposition 8). Using PDOI with $\rho_n = \left(\frac{3}{2}, \frac{3}{2}\right)$ and $w = 1$, we can easily see that every unitary ω-regular representation π of G containing either of these two lowest K-types has infinitesimal character γ_ω.

The K-type $\left(\frac{1}{2}, \frac{1}{2}\right)$ is one-dimensional. In this case, the uniqueness of π follows directly from a result of Zhu [**20**] which states that a representation with scalar lowest K-type is uniquely determined by its infinitesimal character.

It remains to show that there is a unique irreducible representation with lowest K-type $\mu = \left(\frac{3}{2}, \frac{1}{2}\right)$ and infinitesimal character γ_ω. We compute the Vogan-parameter λ_a associated to μ. We have

$$\mu + 2\rho_c = \left(\frac{5}{2}, -\frac{1}{2}\right); \tag{4.74}$$

we choose $\rho = (2, -1)$ to get

$$\mu + 2\rho_c - \rho = \left(\frac{1}{2}, \frac{1}{2}\right). \tag{4.75}$$

This parameter is not in the Weyl chamber determined by ρ, so we project it and obtain

$$\lambda_a = \left(\frac{1}{2}, 0\right). \tag{4.76}$$

So the corresponding Levi factor is

$$MA = Mp(2) \times \widetilde{GL}(1, \mathbb{R}), \tag{4.77}$$

with the discrete series with Harish-Chandra parameter $\frac{1}{2}$ on the first factor, and a character $\chi_{\varepsilon,\nu}$ on the second. In order to obtain infinitesimal character γ_ω on the induced representation, we must have $\nu = \frac{3}{2}$, and the sign ε is uniquely determined by the lowest K-type. So there is indeed only one such representation, which must then be the odd oscillator representation.

4.4.2. *The Sets* Λ_2 *and* Λ_3. Since the elements of these two sets are lowest K-types of $A_{\mathfrak{q}}(\lambda)$ representations with L compact, they are lowest K-types of discrete series. In this case, the representation is determined uniquely and there is nothing to prove.

4.4.3. *The Set* Λ_4. Now suppose that $\mu \in \Lambda_4$. Then

$$\mu = (a, -a) \text{ with } a \geq \frac{3}{2}. \tag{4.78}$$

The corresponding Vogan parameter can easily be computed:

$$\lambda_a = \left(a - \frac{1}{2}, -a + \frac{1}{2}\right), \tag{4.79}$$

so we get $MA = \widetilde{GL}(2,\mathbb{R})$. The representation depends only on a continuous parameter $\nu \geq 0$, and has infinitesimal character

$$\gamma = \left(a - \frac{1}{2} + \nu, -a + \frac{1}{2} + \nu\right). \tag{4.80}$$

Notice that for γ to be ω-regular, we must have $\nu \geq \frac{1}{2}$. Set $\rho_n = \left(\frac{3}{2}, -\frac{1}{2}\right)$; then

$$\mu - \rho_n + \rho_c = (a-1, -a) \tag{4.81}$$

and PDOI gives

$$\langle \gamma, \gamma \rangle \leq a^2 + (a-1)^2. \tag{4.82}$$

It is now easy to check that this condition implies $\nu \leq \frac{1}{2}$. Hence we must have $\nu = \frac{1}{2}$, and we are done with this case.

4.4.4. *The Set* Λ_5. Now consider K-types of the form $\mu = \left(a, \frac{3}{2}\right)$ and $\mu = \left(a, \frac{1}{2}\right)$ with $a \geq \frac{5}{2}$. In both cases, the Vogan algorithm yields

$$\lambda_a = (a-1, 0), \tag{4.83}$$

so the corresponding Levi subgroup is $MA = Mp(2) \times \widetilde{GL}(1,\mathbb{R})$. The representation of $Mp(2)$ is the discrete series representation with Harish-Chandra parameter $a-1$; on $\widetilde{GL}(1,\mathbb{R})$ we have a character $\chi_{\varepsilon,\nu}$. The sign ε is uniquely determined by the lowest K-type ($\mu = \left(a, \frac{3}{2}\right)$ or $\mu = \left(a, \frac{1}{2}\right)$), hence any representation with lowest K-type μ is uniquely determined by its infinitesimal character

$$\gamma = (a-1, \nu). \tag{4.84}$$

It remains to check that the unitarity and ω-regularity conditions determine ν uniquely. Note that in order for γ to be ω-regular, we must have $\nu \geq \frac{1}{2}$.
Now using $\rho_n = \left(\frac{3}{2}, \frac{3}{2}\right)$ for $\mu = \left(a, \frac{3}{2}\right)$, and $\rho_n = \left(\frac{3}{2}, -\frac{1}{2}\right)$ for $\mu = \left(a, \frac{1}{2}\right)$, we obtain

$$\mu - \rho_n + \rho_c = \left(a - 1, \mp\frac{1}{2}\right). \tag{4.85}$$

By PDOI, γ must have length no greater than this parameter, so $\nu \leq \frac{1}{2}$. This forces $\nu = \frac{1}{2}$, proving the uniqueness.

4.5. The Nongenuine Case. In this section, we show that any nongenuine ω-regular unitary representation of $Mp(4)$ is a Meta-$A_{\mathfrak{q}}(\lambda)$ representation.

4.5.1. *Nongenuine K-types.* As for the genuine case, we partition the nongenuine K-types into families, and consider each family in turn. Note that nongenuine K-types are irreducible representations of $U(2)$, parameterized by pairs of integers. Since the calculations are very similar to those performed in the genuine case, we omit many of the details. Partitioning the lowest K-types of Meta-$A_{\mathfrak{q}}(\lambda)$ representations according to the Levi factors L_1 through L_5 of possible parabolic subalgebras as in Section 4.2, we get the families

$$\Lambda_1 = \{(0,0)\}\,, \tag{4.86}$$

$$\Lambda_2 = \{(r,s) : r \geq s \geq 3\}\,, \tag{4.87}$$

$$\Lambda_3 = \{(r.s) : s \leq -1,\, r \geq -s+2\}\,, \tag{4.88}$$

$$\Lambda_4 = \{(r,-r) : r \geq 2\}\,, \tag{4.89}$$

$$\Lambda_5 = \{(r,1) : r \geq 3\}\,. \tag{4.90}$$

The nongenuine K-types which are not lowest K-types of Meta-$A_{\mathfrak{q}}(\lambda)$ representations are therefore part of the set

$$\Sigma = \{(r,2) : r \geq 2\} \cup \{(r,0) : r \geq 1\} \cup \{(r+1,-r) : r \geq 1\} \cup \{(2,1)\,, (1,1)\,, (1,-1)\}\,.$$

4.5.2. *The K-Types in Σ.* For every K-type μ in the set Σ we can show, using PDOI, that there is no ω-regular unitary representation which has μ as one of its lowest K-types. For example, for

$$\mu = (2,1) \tag{4.91}$$

we can use $\rho_n = \left(\frac{3}{2}, \frac{3}{2}\right)$ to see that

$$\mu - \rho_n + \rho_c = (1,-1)\,, \tag{4.92}$$

a weight whose length is strictly less than that of any ω-regular infinitesimal character. Because PDOI fails, an ω-regular representation with lowest K-type μ cannot be unitary. For a second example, let's consider

$$\mu = (r,0) \tag{4.93}$$

with $r \geq 1$. The corresponding Vogan-parameter is

$$\lambda_a = (r-1,0)\,. \tag{4.94}$$

If $r = 1$, then $\lambda_a = (0,0)$, so any representation with lowest K-type $\mu = (r,0)$ is a principal series. In this case, with $\rho_n = \left(\frac{3}{2}, -\frac{1}{2}\right)$ and w the long element of the Weyl group, we get

$$w(\mu - \rho_n) + \rho_c = (1,-1)\,, \tag{4.95}$$

once again a weight whose length is less than that of γ_ω.
If $r \geq 2$, then the corresponding Levi subgroup is $MA = Mp(2) \times \widetilde{GL}(1,\mathbb{R})$, and the infinitesimal character is

$$\gamma = (r-1,\nu)\,. \tag{4.96}$$

The ω-regularity condition forces $\nu \geq \frac{1}{2}$. If we choose $\rho_n = \left(\frac{3}{2}, -\frac{1}{2}\right)$ and $w = 1$, we get

$$w(\mu - \rho_n) + \rho_c = (r-1,0)\,, \tag{4.97}$$

a weight of length strictly smaller than γ. In either case, we conclude there is no ω-regular unitary representation with lowest K-type $(r,0)$.

The calculation for the remaining K-types in Σ is similar; we leave the details to the diligent reader.

4.5.3. *The Lowest K-Types of Meta-$A_{\mathfrak{q}}(\lambda)$'s.* As for the genuine case, it remains to show that every ω-regular unitary representation with a lowest K-type of a Meta-$A_{\mathfrak{q}}(\lambda)$ is indeed a Meta-$A_{\mathfrak{q}}(\lambda)$. The proof for K-types in Λ_2, Λ_3, and Λ_4 is very similar to the corresponding one in the genuine case, so we omit it.

For $\mu = (0,0)$, we need to show that the trivial representation is the only ω-regular unitary spherical representation of $Mp(4)$. Here, we refer to the results of [**3**] or [**10**].

We are left with the K-types of the form

$$\mu = (r,1) \text{ with } r \geq 3. \tag{4.98}$$

These are lowest K-types of Meta-$A_{\mathfrak{q}}(\lambda)$ representations $\mathcal{R}_{\mathfrak{q}}(\mathbb{C}_\lambda \otimes J_\nu)$ (cf. Definition 2.5) with $L = L_5 = \widetilde{U}(1,0) \times Mp(2)$, $\lambda = r-3$, and J_ν a spherical complementary series of $SL(2,\mathbb{R})$ with infinitesimal character $\frac{1}{2} \leq \nu \leq 1$. Such a representation has infinitesimal character

$$(\lambda,\nu) + \rho(\mathfrak{u}) = (r-3,\nu) + (2,0) = (r-1,\nu)\,. \tag{4.99}$$

We prove that every ω-regular unitary representation X with lowest K-type $\mu = (r,1)$ is of the form $\mathcal{R}_{\mathfrak{q}}(\mathbb{C}_\lambda \otimes J_\nu)$, with $\lambda = r-3$ and $\frac{1}{2} \leq \nu \leq 1$.
The Vogan parameter associated to μ is

$$\lambda_a = (r-1,0) \tag{4.100}$$

as in (4.94) above. Hence X has infinitesimal character

$$\gamma = (r-1,\nu) \tag{4.101}$$

for some positive number ν. The representation X is uniquely determined by the value of ν, so must be of the form $\mathcal{R}_{\mathfrak{q}}(\mathbb{C}_\lambda \otimes J_\nu)$, with $\lambda = r-3$. It remains to prove that ν belongs to the appropriate range. Applying PDOI to μ with $\rho_n = \left(\frac{3}{2},-\frac{1}{2}\right)$, we find that the infinitesimal character γ of X can not be greater than

$$(r-1,1)\,. \tag{4.102}$$

Therefore $\nu \leq 1$. The ω-regularity condition forces $\nu \geq \frac{1}{2}$, so we are done.

Consequently, we have

PROPOSITION 10. If X is an irreducible ω-regular unitary nongenuine representation of $Mp(4)$ then X is a Meta-$A_{\mathfrak{q}}(\lambda)$ representation.

5. Nonunitarity of the Mystery Representation X_M

In this section we finish the proof of Theorem 2.8 by showing that the two representations of $Mp(4)$ with lowest K-type $\left(\frac{1}{2},-\frac{1}{2}\right)$ and infinitesimal character $\left(\frac{3}{2},\frac{1}{2}\right)$ are not unitary (cf. Section 4.3). Since the PDOI is satisfied in this case, we have to use a different method; i. e., we must construct the intertwining operator giving rise to the invariant hermitian form and show that the form is indefinite. We discuss this theory in Section 5.2 in some detail.

5.1. The Representation X_M. We consider the irreducible representations of $Mp(4)$ with lowest K-type $\left(\frac{1}{2}, -\frac{1}{2}\right)$ and infinitesimal character $\left(\frac{3}{2}, \frac{1}{2}\right)$, and prove that they are nonunitary. From now on we identify elements of K with their images under the map ι (cf. (4.3)).

The K-type $\left(\frac{1}{2}, -\frac{1}{2}\right)$ is fine, so any representation with this lowest K-type is a constituent of a principal series, i. e., an induced representation of the form

$$I_P(\delta, \nu) = \mathrm{Ind}_P^G(\delta \otimes \nu \otimes 1), \tag{5.1}$$

where $P = MAN$ is a minimal parabolic subgroup of G, M and A are as in Section 4.1, δ is a character of M, and ν is the character of A with differential $\left(\frac{3}{2}, \frac{1}{2}\right)$. Recall that, by Frobenius reciprocity, the K-type

$$\mu_\delta = \left(\frac{1}{2}, -\frac{1}{2}\right) \tag{5.2}$$

occurs in the principal series $I_P(\delta, \nu)$ if (and only if) the character δ is a summand of the restriction of μ_δ to M. We look at this restriction.
The K-type μ_δ contains the weights

$$\left(\frac{1}{2}, -\frac{1}{2}\right) \text{ and } \left(-\frac{1}{2}, \frac{1}{2}\right); \tag{5.3}$$

they correspond to the M-characters

$$\delta_1(a) = +i = -\delta_1(b) \tag{5.4}$$

$$\delta_2(a) = -i = -\delta_2(b) \tag{5.5}$$

respectively. The two elements

$$a = \left[\begin{pmatrix} -1 & 0 \\ 0 & 1 \end{pmatrix}, i\right] \text{ and } b = \left[\begin{pmatrix} 1 & 0 \\ 0 & -1 \end{pmatrix}, i\right] \tag{5.6}$$

are the generators of M. Then $I_P(\delta, \nu)$ contains μ_δ if and only if δ equals δ_1 or δ_2. We conclude that there are exactly two irreducible representations of $Mp(4)$ with lowest K-type μ_δ and infinitesimal character $\left(\frac{3}{2}, \frac{1}{2}\right)$: they are the irreducible constituents of the principal series $I_P(\delta_1, \nu)$ and $I_P(\delta_2, \nu)$ containing μ_δ. We denote these representations by $\overline{X}(\delta_1, \nu)$ and $\overline{X}(\delta_2, \nu)$.

It is not hard to see that $\overline{X}(\delta_1, \nu)$ and $\overline{X}(\delta_2, \nu)$ are contragredient of each other (hence one is unitary if and only if the other one is). We sometimes refer to either one of these two representations as the "Mystery Representation" X_M, and write δ for either one of the two characters δ_i.

The purpose of this section is to prove that the Mystery representation is nonunitary. First we observe that X_M is Hermitian (i. e., it is equivalent to its Hermitian dual and hence carries a nondegenerate G-invariant Hermitian form). By [**17**], it is sufficient to prove that there exists an element w of the Weyl group $W(G, A)$ taking (P, δ, ν) to $\left(\bar{P}, \delta, -\overline{\nu}\right)$, with $\bar{P}$ the opposite parabolic. Note that $-\overline{\nu} = -\nu$ because ν is real.
We claim that the long Weyl group element $w_0 = s_{2e_1} s_{2e_2}$ has the desired properties. Indeed, w_0 takes P to its opposite, ν to its negative, and fixes δ:

$$(w_0 \cdot \delta)(m) = \delta(\sigma_0^{-1} m \sigma_0) = \delta(m) \quad \forall m \in M. \tag{5.7}$$

Here σ_0 denotes a representative for w_0 in K:

$$\sigma_0 = \sigma_{2e_1}\sigma_{2e_2} = \left[\begin{pmatrix} i & 0 \\ 0 & 1 \end{pmatrix}, e^{\frac{\pi}{4}i}\right]\left[\begin{pmatrix} 1 & 0 \\ 0 & i \end{pmatrix}, e^{\frac{\pi}{4}i}\right] = \left[\begin{pmatrix} i & 0 \\ 0 & i \end{pmatrix}, i\right] \tag{5.8}$$

(note that σ_0 is diagonal, hence it commutes with all elements of M).

The nondegenerate invariant Hermitian form on X_M is unique up to a constant. We will explicitly construct a Hermitian intertwining operator that induces the form. Proving that X_M is nonunitary is then equivalent to showing that this intertwining operator is indefinite.

5.2. Intertwining Operators and Unitarity of Principal Series. In this section, we review the theory of intertwining operators for principal series of (double covers of) real split groups. The results presented at the beginning of the section are well known in the literature (c.f. [**5**], [**16**], [**2**] and [**7**]) and are reported here for completeness. The content of the last part is more innovative. The idea, explained at the end of the section, of using the operator $l_\mu(w_0, \nu)$ to look at several principal series at the same time, is due to Barbasch and Pantano, and has not appeared in any published work yet.

5.2.1. *(Formal) Intertwining Operators for Principal Series.* Let G be a (possibly trivial) two-fold cover of the split real form of a connected (simple and simply connected) reductive algebraic group. By allowing the cover to be trivial, we intend to discuss split linear groups and their nonlinear double cover simultaneously.

Choose a minimal parabolic subgroup $P = MAN$ of G. For every irreducible representation (δ, V^δ) of M and every character ν of A, we write

$$I_P(\delta, \nu) = \mathrm{Ind}_P^G(\delta \otimes \nu \otimes 1)$$

for the induced representation of G. This is normalized induction, so $I_P(\delta, \nu)$ is unitary when $\delta \otimes \nu \otimes 1$ is. The representation space for $I_P(\delta, \nu)$ is denoted by $\mathcal{H}_P(\delta, \nu)$, and consists of functions

$$f \colon G \to V^\delta \tag{5.9}$$

whose restriction to K is square integrable, such that

$$f(gman) = e^{-(\nu+\rho)\log a}\delta(m)^{-1} f(g) \tag{5.10}$$

for all $g \in G$, and $man \in P$. The action of G on these functions is by left translation.

For every element w of the Weyl group, there is a *formal* intertwining operator

$$\mathcal{A}_P(w, \delta, \nu) \colon I_P(\delta, \nu) \to I_P(w\delta, w\nu)\,. \tag{5.11}$$

Notice that $\mathcal{A}_P(w, \delta, \nu)$ satisfies all the intertwining properties, but is defined by an integral that may not converge for all values of ν. (See [**5**] for details.)
For all $F \colon G \to V^\delta$ in $\mathcal{H}_P(\delta, \nu)$, we set:

$$\mathcal{A}_P(w, \delta, \nu) f \colon G \to V^{w\delta},\ g \mapsto \int_{\overline{N} \cap (wNw^{-1})} f(gw\overline{n})\, d\overline{n}. \tag{5.12}$$

To reduce the computation of $\mathcal{A}_P(w, \delta, \nu)$ to a finite-dimensional problem, we restrict the operator to the various K-types appearing in the principal series. For every $(\mu, E_\mu) \in \widehat{K}$, we obtain an operator

$$\widetilde{a}_\mu(w, \delta, \nu) \colon \mathrm{Hom}_K(\mu, \mathrm{Ind}_P(\delta, \nu)) \to \mathrm{Hom}_K(\mu, \mathrm{Ind}_P(w\delta, w\nu)) \tag{5.13}$$

by composition on the range. Note that the restriction of $\mathrm{Ind}_P(\delta, \nu)$ to K is independent of ν and equal to the induced representation $\mathrm{Ind}_M^K \delta$. Similarly, $\mathrm{Ind}_P(w\delta, w\nu)$ restricts to $\mathrm{Ind}_M^K w\delta$. Then we can interpret $\widetilde{a}_\mu(w, \delta, \nu)$ as an operator

$$\widetilde{a}_\mu(w, \delta, \nu) \colon \mathrm{Hom}_K(\mu, \mathrm{Ind}_M^K \delta) \to \mathrm{Hom}_K(\mu, \mathrm{Ind}_M^K w\delta). \tag{5.14}$$

By Frobenius reciprocity, we obtain an operator

$$a_\mu(w,\delta,\nu)\colon \operatorname{Hom}_M(\mu,\delta)\to \operatorname{Hom}_M(\mu,w\delta) \tag{5.15}$$

for every $\mu\in\widehat{K}$. An easy computation shows that

$$(a_\mu(w,\delta,\nu)T)(v)=\int_{\overline{N}\cap(\sigma N\sigma^{-1})} e^{-(\nu+\rho)\log \underline{a}(\overline{n})}T\left(\mu\left(\sigma\underline{k}\left(\overline{n}\right)\right)^{-1}v\right)d\overline{n} \tag{5.16}$$

for all $T\in\operatorname{Hom}_M(\mu,\delta)$ and all $v\in E_\mu$. (See [7] for details.) Here σ denotes a representative for w in K, and $\underline{k}(g)\underline{a}(g)\underline{n}(g)$ denotes the Iwasawa decomposition of an element g in $G=KAN$.

We are going to break down the operator $a_\mu(w,\delta,\nu)$ so that its computation becomes manageable. The factorization

$$\mathcal{A}_P(w_1w_2,\delta,\nu)=\mathcal{A}_P(w_1,w_2\delta,w_2\nu)\mathcal{A}_P(w_2,\delta,\nu) \tag{5.17}$$

holds for any pair of Weyl group elements satisfying the condition

$$l(w_1w_2)=l(w_1)+l(w_2).$$

Here l denotes the length function on W. It follows that any formal operator $\mathcal{A}_P(w,\delta,\nu)$ can be decomposed as a product of operators corresponding to simple root reflections. The operators $a_\mu(w,\delta,\nu)$ inherit a similar decomposition. In particular, if

$$w=s_{\alpha_r}s_{\alpha_{r-1}}\cdots s_{\alpha_1} \tag{5.18}$$

is a *minimal* decomposition of w as a product of simple reflections, then we can factor $a_\mu(w,\delta,\nu)$ as a product of operators of the form

$$a_\mu(s_{\alpha_i},\delta_{i-1},\nu_{i-1})\colon \operatorname{Hom}_M(\mu,\delta_{i-1})\to \operatorname{Hom}_M(\mu,s_{\alpha_i}\delta_{i-1}) \tag{5.19}$$

for α_i a simple root, $\delta_{i-1}\in\widehat{M}$ (in the W-orbit of δ) and $\nu_{i-1}\in\mathfrak{a}^*$.

In view of this result, we only need to understand the operator $a_\mu(s_\alpha,\delta,\nu)$ for α simple. The computation of $a_\mu(s_\alpha,\delta,\nu)$ can largely be reduced to a similar computation for the rank-one group MG^α, where G^α is the $SL(2)$ or $Mp(2)$ subgroup of G attached to the root α. (See Section 5.3 for a description of G^α for $Mp(4)$.)

We recall the construction of the group G^α. For every (simple) root α we can define a Lie algebra homomorphism

$$\phi_\alpha\colon \mathfrak{sl}(2,\mathbb{R})\to\mathfrak{g}_0 \tag{5.20}$$

as in (4.3.5) of [16]. The image of ϕ_α is a subalgebra of $\mathfrak{g}_0$ isomorphic to $\mathfrak{sl}(2,\mathbb{R})$. The corresponding connected subgroup of G is denoted by G^α.

Exponentiating ϕ_α, we obtain a group homomorphism

$$\Phi_\alpha\colon Mp(2,\mathbb{R})\to G \tag{5.21}$$

with image G^α. The structure of G^α depends on whether the map Φ_α factors through $SL(2,\mathbb{R})$.

If Φ_α factors through $SL(2,\mathbb{R})$, we say that the root α is "not metaplectic". In this case, G^α is isomorphic to $SL(2,\mathbb{R})$, and the maximal compact subgroup $K^\alpha\subset G^\alpha$ is isomorphic to $U(1)$. If Φ_α does not factor through $SL(2,\mathbb{R})$, then we call α "metaplectic". For metaplectic roots, $G^\alpha\simeq Mp(2,\mathbb{R})$ and $K^\alpha\simeq\widetilde{U(1)}$. We can give a more explicit description of the metaplectic roots: if G is not of type G_2, then the metaplectic roots are exactly the long roots. If G is of type G_2, then all roots are metaplectic.

Let us go back to the task of computing the operator $a_\mu(s_\alpha, \delta, \nu)$. Recall that

$$a_\mu(s_\alpha, \delta, \nu)T(v) = \int_{\bar{N}\cap(\sigma_\alpha N \sigma_\alpha^{-1})} e^{-(\nu+\rho)\log \underline{a}(\bar{n})} T\left(\mu\left(\sigma_\alpha \underline{k}(\bar{n})\right)^{-1} v\right) d\bar{n}$$

for all $T \in \mathrm{Hom}_M(\mu, \delta)$ and all $v \in E_\mu$. Let $G^\alpha = K^\alpha A^\alpha N^\alpha$ be the Iwasawa decomposition of G^α. Note that:

- $\bar{N} \cap (\sigma_\alpha N \sigma_\alpha^{-1}) = \bar{N}^\alpha$
- The Iwasawa decompositions of $\bar{n} \in \bar{N}^\alpha$ inside G and G^α coincide, because
$$K^\alpha = K \cap G^\alpha \qquad A^\alpha = A \cap G^\alpha \qquad N^\alpha = N \cap G^\alpha.$$
- If α is simple, the restriction of $\rho = \frac{1}{2}\left[\sum_{\alpha \in \Delta^+} \alpha\right]$ to $\mathrm{Lie}(A^\alpha)$ equals ρ^α.

Then we can write

$$a_\mu(s_\alpha, \delta, \nu)T(v) \int_{\bar{N}^\alpha} e^{-(\nu+\rho^\alpha)\log \underline{a}^\alpha(\bar{n})} T\left(\mu\left(\sigma_\alpha \underline{k}^\alpha(\bar{n})\right)^{-1} v\right) d\bar{n} \tag{5.22}$$

for all $T \in \mathrm{Hom}_M(\mu, \delta)$ and $v \in E_\mu$. This formula coincides with the one for the corresponding operator for the group MG^α on the representation $\mu|_{MK^\alpha}$.

It follows that any decomposition of μ into MK^α-invariant subspaces must be preserved by the operator $a_\mu(s_\alpha, \delta, \nu)$.

Let $\mu = \bigoplus_{l \in \mathbb{Z}/2} \varphi_l^\alpha$ be the decomposition of μ as a direct sum of isotypic components of characters of K^α. The group M stabilizes the subspace $(\varphi_l^\alpha + \varphi_{-l}^\alpha)$, for every $l \in \mathbb{N}/2$. Hence the decomposition

$$\mu = \bigoplus_{l \in \mathbb{N}/2} (\varphi_l^\alpha + \varphi_{-l}^\alpha) \tag{5.23}$$

is stable under the action of MK^α. The corresponding decompositions of $\mathrm{Hom}_M(\mu, \delta)$ and $\mathrm{Hom}_M(\mu, s_\alpha\delta)$ are stable under the action of $a_\mu(s_\alpha, \delta, \nu)$. More precisely,

$$a_\mu(s_\alpha, \delta, \nu)\colon \mathrm{Hom}_M\left(\varphi_l^\alpha + \varphi_{-l}^\alpha, \delta\right) \to \mathrm{Hom}_M\left(\varphi_l^\alpha + \varphi_{-l}^\alpha, s_\alpha\delta\right) \tag{5.24}$$

for all $l \in \mathbb{N}/2$.

The computation of $a_\mu(s_\alpha, \delta, \nu)$ on $\mathrm{Hom}_M\left(\varphi_l \oplus \varphi_{-l}, V^\delta\right)$ can be carried out explicitly, by evaluating an integral in $SL(2, \mathbb{R})$ or $Mp(2)$. We will not display the calculation here but just state the result. If

$$T \in \mathrm{Hom}_M(\varphi_l \oplus \varphi_{-l}, V^\delta),$$

then $a_\mu(s_\alpha, \delta, \nu)$ maps T to

$$c_l(\alpha, \nu)\, T \circ \mu(\sigma_\alpha^{-1}). \tag{5.25}$$

The constant $c_l(\alpha, \nu)$ depends on l and on the inner product $\langle \nu, \alpha^\vee \rangle$, and is equal to:

$$c_l(\alpha, \nu) := \frac{\pi\Gamma(\lambda)}{2^{\lambda-1}\Gamma\left(1 + \frac{\lambda-1+l}{2}\right)\Gamma\left(1 + \frac{\lambda-1-l}{2}\right)}. \tag{5.26}$$

Here $\lambda = \langle \nu, \check{\alpha} \rangle$, and Γ denotes the Gamma function. Note that $c_l(\alpha, \nu) = c_{-l}(\alpha, \nu)$.

To simplify the notation, we introduce a normalization.

Choose a fine K-type μ_δ containing δ. The space $\mathrm{Hom}_M(\mu_\delta, \delta)$ is one-dimensional, hence the operator $a_{\mu_\delta}(w, \delta, \nu)$ acts on it by a scalar. We normalize the operator $\mathcal{A}_P(w, \delta, \nu)$ so that this scalar is one.

On each K-type μ we obtain a normalized operator $a'_\mu(s_\alpha,\delta,\nu)$, which acts on $\mathrm{Hom}_M(\varphi^\alpha_l+\varphi^\alpha_{-l},\delta)$ by

$$a'_\mu(s_\alpha,\delta,\nu)T = c'_l(\alpha,\nu)\,T\circ\mu(\sigma_\alpha^{-1})$$

where

$$c'_l(\alpha,\nu)=\frac{c_l(\alpha,\nu)}{c_{\frac12}(\alpha,\nu)} \qquad \text{if } l \text{ belongs to } \mathbb{Z}+\frac{1}{2};$$

$$c'_l(\alpha,\nu)=\frac{c_l(\alpha,\nu)}{c_1(\alpha,\nu)} \qquad \text{if } l \text{ is an odd integer;}$$

$$c'_l(\alpha,\nu)=\frac{c_l(\alpha,\nu)}{c_0(\alpha,\nu)} \qquad \text{if } l \text{ is an even integer.}$$

Using the expression of $c_l(\alpha,\nu)$ in terms of Γ functions, and the property

$$\Gamma(z+n)=z(z+1)(z+2)\cdots(z+n-1)\Gamma(z) \quad \forall n>0$$

of the Γ function, we obtain:

- $c'_{-\frac12+2m}(\alpha,\nu)=(-1)^m\frac{(\frac12-\lambda)(\frac52-\lambda)\cdots(2m-\frac32-\lambda)}{(\frac12+\lambda)(\frac52+\lambda)\cdots(2m-\frac32+\lambda)}$
- $c'_{\frac12+2m}(\alpha,\nu)=(-1)^m\frac{(\frac32-\lambda)(\frac72-\lambda)\cdots(2m-\frac12-\lambda)}{(\frac32+\lambda)(\frac72+\lambda)\cdots(2m-\frac12+\lambda)}$
- $c'_{2m+1}(\alpha,\nu)=(-1)^m\frac{(2-\lambda)(4-\lambda)\cdots(2m-\lambda)}{(2+\lambda)(4+\lambda)\cdots(2m+\lambda)}$
- $c'_{2m}(\alpha,\nu)=(-1)^m\frac{(1-\lambda)(3-\lambda)\cdots(2m-1-\lambda)}{(1+\lambda)(3+\lambda)\cdots(2m-1+\lambda)}$

for all $m>0$, and of course $c'_{\frac12}(\alpha,\nu)=c'_1(\alpha,\nu)=c'_0(\alpha,\nu)=1$. As usual, $\lambda=\langle\nu,\check\alpha\rangle$.

From now on, all our intertwining operators will be normalized. With abuse of notation, we will replace $a'_\mu(w,\delta,\nu)$ by $a_\mu(w,\delta,\nu)$ and $c'_l(\alpha,\nu)$ by $c_l(\alpha,\nu)$. In particular, we will write:

$$c_l(\alpha,\nu)=\begin{cases}1 & \text{if } l=0,1,\text{or } \frac12\\[2ex] -\frac{\frac12-\lambda}{\frac12+\lambda}=-\frac{\frac12-\langle\nu,\alpha^\vee\rangle}{\frac12+\langle\nu,\alpha^\vee\rangle} & \text{if } l=\frac32\\[2ex] -\frac{1-\lambda}{1+\lambda}=-\frac{1-\langle\nu,\alpha^\vee\rangle}{1+\langle\nu,\alpha^\vee\rangle} & \text{if } l=2.\end{cases} \tag{5.27}$$

The action of the operator

$$a_\mu(s_\alpha,\delta,\nu)\colon \mathrm{Hom}_M(\mu,\delta)\to\mathrm{Hom}_M(\mu,s_\alpha\delta)$$

is now completely understood. To conclude the subsection, we specify the parity of the indices l appearing in the decomposition

$$\mathrm{Hom}_M(\mu,\delta)=\bigoplus_{l\in\mathbb{N}/2}\mathrm{Hom}_M(\varphi^\alpha_l+\varphi^\alpha_{-l},\delta).$$

The Lie algebra of K^α is generated by the element

$$Z_\alpha:=\phi_\alpha\begin{pmatrix}0&1\\-1&0\end{pmatrix}.$$

Because $K^\alpha = \exp(i\mathbb{R}Z_\alpha)$, we can identify the space ϕ_l^α (defined to be the isotypic component of the character l of K^α inside μ) with the l-generalized eigenspace of $d\mu(iZ_\alpha)$. We note that the element $m_\alpha = \exp(i\pi Z_\alpha)$ has order 2 if α is not metaplectic, and has order 4 otherwise. This condition imposes strong restrictions on the possible eigenvalues of $d\mu(iZ_\alpha)$.
If α is not metaplectic, then $d\mu(iZ_\alpha)$ has integer eigenvalues for every K-type μ. If α is metaplectic, then the eigenvalues of $d\mu(iZ_\alpha)$ are integers if the K-type μ is non-genuine (i.e. if $\mu(-I)$ is trivial), and half-integers if μ is genuine. Therefore:

- if μ is genuine and α is metaplectic, then every index l appearing in the decomposition
$$\mathrm{Hom}_M(\mu,\delta) = \bigoplus_{l\in\mathbb{N}/2} \mathrm{Hom}_M(\varphi_l^\alpha + \varphi_{-l}^\alpha, \delta)$$
belongs to $\mathbb{Z} + \frac{1}{2}$;
- if μ is not genuine, or if μ is genuine and α is not metaplectic, then every index l belongs to $\mathbb{Z}$. Precisely, l is an odd integer if $\delta(m_\alpha) = -I$ and it is an even integer otherwise.

Finally, we observe that the action of Z_α on E_μ extends to an action of Z_α^2 on the space $\mathrm{Hom}_M(\mu,\delta)$, because Z_α^2 commutes with M. The action is given by:
$$T \mapsto T \circ (d\mu(Z_\alpha))^2 .$$
So we can think of
$$\mathrm{Hom}_M(\mu,\delta) = \bigoplus_{l\in\mathbb{N}/2} \mathrm{Hom}_M(\varphi_l^\alpha + \varphi_{-l}^\alpha, \delta)$$
as the decomposition of $\mathrm{Hom}_M(\mu,\delta)$ into generalized eigenspaces for Z_α^2.

5.2.2. *Unitarity of Langlands quotients.* We restrict our attention to representations with real infinitesimal character, so we assume ν to be real.
Suppose that ν is strictly dominant for the positive root system determined by N. Then the principal series $I_P(\delta,\nu)$ has a unique irreducible Langlands quotient $\bar{X}_P(\delta,\nu)$. The purpose of this section is to discuss the unitarity of $\bar{X}_P(\delta,\nu)$.

By work of Knapp and Zuckermann, $\bar{X}_P(\delta,\nu)$ is Hermitian if and only if the long Weyl group element w_0 satisfies
$$w_0\delta \simeq \delta \quad \text{and} \quad w_0\nu = -\nu.$$
We will assume that the above conditions are met. Then the normalized operator $\mathcal{A}_P(w_0,\delta,\nu)$ (introduced in the previous section) is defined by a converging integral. Let $\delta(w_0)$ be any intertwining operator displaying $w_0\delta$ as equivalent to δ. The composition
$$L(w_0,\delta,\nu) := \delta(w_0)\mathcal{A}_P(w_0,\delta,\nu)\colon I_P(\delta,\nu) \to I_P(\delta,-\nu) \tag{5.28}$$
is Hermitian, and induces a G-invariant form $\langle\,,\rangle$ on $\mathcal{H}_P(\delta,\nu)$ by
$$\langle f,g\rangle = (L(w_0,\delta,\nu)f,g)_{L^2(K)} \qquad \forall\, f,g \in \mathcal{H}_P(\delta,\nu). \tag{5.29}$$
Here
$$(f,g)_{L^2(K)} = \int_K f(k)\overline{g(k)}dk. \tag{5.30}$$
The form $\langle\,,\rangle$ descends to a nondegenerate G-invariant form on the quotient of $\mathcal{H}_P(\delta,\nu)$ by the kernel of the operator $\mathcal{A}_P(w_0,\delta,\nu)$, which is isomorphic to $\bar{X}_P(\delta,\nu)$.

Note that the irreducible Hermitian representation $\bar{X}_P(\delta,\nu)$ admits a unique nondegenerate invariant form (up to a constant); hence proving that $\bar{X}_P(\delta,\nu)$ is nonunitary amounts to showing that the form $\langle\, ,\rangle$ is indefinite.

We begin by making a convenient choice of $\delta(w_0)$. Let $(\mu_\delta, E_{\mu_\delta})$ be a fine K-type containing δ. Because μ_δ contains δ with multiplicity one, we can canonically identify δ with its unique copy inside μ_δ. Then V^δ is identified with the isotypic component $E_{\mu_\delta}(\delta)$ of δ in E_{μ_δ}. (Recall that V^δ is the representation space for δ.) We define two actions of M on V^δ: one is the restriction of μ_δ to M (identified with δ), the other is by $w_0\cdot\delta$. The operator $\mu_\delta(w_0)$ maps $V^\delta = E_{\mu_\delta}(\delta)$ into itself (because $w_0\cdot\delta \simeq \delta$), and intertwines the two actions. Hence we can choose $\delta(w_0) = \mu_\delta(w_0)$.

This gives:

$$L(w_0,\delta,\nu) = \mu_\delta(w_0)\mathcal{A}_P(w_0,\delta,\nu). \tag{5.31}$$

By Frobenius reciprocity, $L(w_0,\delta,\nu)$ gives rise to an operator

$$l_\mu(w_0,\delta,\nu)\colon \operatorname{Hom}_M(\mu,\delta) \to \operatorname{Hom}_M(\mu,\delta) \tag{5.32}$$

for every K-type μ appearing in the principal series. The operator $l_\mu(w_0,\delta,\nu) := \mu_\delta(w_0)a_\mu(w_0,\delta,\nu)$ carries all the signature information on the K-type μ, and is zero if μ does not appear in the quotient $\bar{X}(\delta,\nu)$.

Next, we give a factorization of $l_\mu(w_0,\delta,\nu)$ as a product of operators corresponding to simple root reflections. Recall that if

$$w_0 = s_{\alpha_r}s_{\alpha_{r-1}}\cdots s_{\alpha_1}$$

is a *minimal* decomposition of w_0 as a product of simple reflections, then the operator $a_\mu(w_0,\delta,\nu)$ factors as

$$a_\mu(w_0,\delta,\nu) = \prod_{j=1}^{r} a_\mu(s_{\alpha_j},\delta_{j-1},\nu_{j-1})$$

with $x_0 = 1$, $\delta_0 = \delta = x_0\cdot\delta$, $\nu_0 = \nu = x_0\cdot\nu$, and

$$\delta_j = \underbrace{s_{\alpha_j}s_{\alpha_{j-1}}\cdots s_{\alpha_1}}_{x_j}\cdot\delta = x_j\cdot\delta \qquad \nu_j = \underbrace{s_{\alpha_j}s_{\alpha_{j-1}}\cdots s_{\alpha_1}}_{x_j}\cdot\nu = x_j\cdot\nu$$

for $j\ge 1$. So we can write:

$$\begin{aligned}
l_\mu(w_0,\delta,\nu) = \mu_\delta(w_0)a_\mu(w_0,\delta,\nu) &= \mu_\delta(w_0)\left[\prod_{j=1}^{r} a_\mu(s_{\alpha_j},\delta_{j-1},\nu_{j-1})\right] = \\
&= \mu_\delta(x_r)\left[\prod_{j=1}^{r} a_\mu(s_{\alpha_j},\delta_{j-1},\nu_{j-1})\right]\mu_\delta(x_0)^{-1} = \\
&= \prod_{j=1}^{r}\left[\mu_\delta(x_j)a_\mu(s_{\alpha_j},\delta_{j-1},\nu_{j-1})\mu_\delta(x_{j-1})^{-1}\right] = \\
&= \prod_{j=1}^{r}\left[\mu_\delta(\sigma_{\alpha_j})a_\mu(s_{\alpha_j},\rho_{j-1},\nu_{j-1})\right] = \prod_{j=1}^{r} l_\mu(s_{\alpha_j},\rho_{j-1},\nu_{j-1})
\end{aligned}$$

The operator $a_\mu(s_{\alpha_j},\delta_{j-1},\nu_{j-1})$ carries $\operatorname{Hom}_M(\mu,\delta_{j-1})$ into $\operatorname{Hom}_M(\mu,\delta_j)$. For all $k = 1\dots r$, the M-type $\delta_k = x_k\delta$ is the representation

$$\delta_k(m)v = \delta(\sigma_k^{-1}m\sigma_k)v$$

of M on the space $V^\delta = E_{\mu_\delta}(\delta)$. Here σ_k is a representative in K for the root reflection x_k. To obtain a more natural realization of the representation δ_k, we will replace δ_k by its unique copy inside μ_δ.
The fine K-type μ_δ contains every M-type in the Weyl group orbit of δ with multiplicity one. In particular, μ_δ contains a unique representation ρ_k isomorphic to δ_k. The representation space for ρ_k is the isotypic component of δ_k in E_{μ_δ}, denoted by $E_{\mu_\delta}(\delta_k)$; the action of M on $E_{\mu_\delta}(\delta_k)$ is given by (the restriction of) μ_δ. The map $\mu_\delta(x_k)$ carries $V^\delta = E_{\mu_\delta}(\delta)$ into $E_{\mu_\delta}(x_k\delta) = E_{\mu_\delta}(\delta_k)$, and intertwines the δ_k-action of M on V^δ with the ρ_k-action of M on $E_{\mu_\delta}(\delta_k)$. Applying $\mu_\delta(x_k)$ to the range, we obtain an isomorphism

$$\mu_\delta(x_k)\colon \operatorname{Hom}_M(\mu, \delta_k) \to \operatorname{Hom}_M(\mu, \rho_k), \tag{5.33}$$

for all $k = 1 \dots n$. Let us return to our factorization. Recall that

$$l_\mu(w_0, \delta, \nu) = \prod_{j=1}^{r} \left[\mu_\delta(x_j) a_\mu(s_{\alpha_j}, \delta_{j-1}, \nu_{j-1}) \mu_\delta(x_{j-1})^{-1}\right].$$

For all $j = 1 \dots r$, there is a commutative diagram:

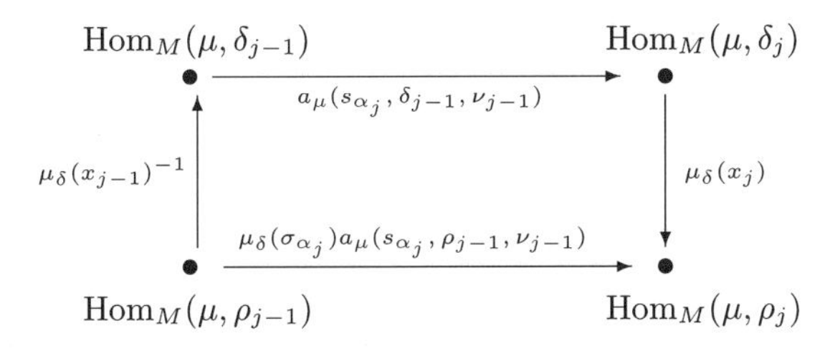

hence we can write

$$l_\mu(w_0, \delta, \nu) = \prod_{j=1}^{r} \left[\mu_\delta(\sigma_{\alpha_j}) a_\mu(s_{\alpha_j}, \rho_{j-1}, \nu_{j-1})\right] = \prod_{j=1}^{r} l_\mu(s_{\alpha_j}, \rho_{j-1}, \nu_{j-1}). \tag{5.34}$$

For all $j = 1 \dots r$, we set $l_\mu(s_{\alpha_j}, \rho_{j-1}, \nu_{j-1}) := \mu_\delta(\sigma_{\alpha_j}) a_\mu(s_{\alpha_j}, \rho_{j-1}, \nu_{j-1})$, and we regard $l_\mu(s_{\alpha_j}, \rho_{j-1}, \nu_{j-1})$ as an operator acting on $\operatorname{Hom}_M(E_\mu, E_{\mu_\delta})$.

The operators $l_\mu(w_0, \delta_k, \nu)$ coming from characters δ_k in the Weyl group orbit of δ admit a similar decomposition. Notice that

- We can combine the various operators $l_\mu(w_0, \delta_k, \nu)$ to get an operator

$$l_\mu(w_0, \nu)\colon \operatorname{Hom}_M(E_\mu, E_{\mu_\delta}) \to \operatorname{Hom}_M(E_\mu, E_{\mu_\delta}). \tag{5.35}$$

- The operator $l_\mu(w_0, \nu)$ factors as a product of operators corresponding to simple reflections:

$$l_\mu(w_0, \nu) = \prod_{j=1}^{r} l_\mu(s_{\alpha_j}, \nu_{j-1}). \tag{5.36}$$

  Each factor can be interpreted as an operator acting on $\operatorname{Hom}_M(E_\mu, E_{\mu_\delta})$, and is easy to compute: write

$$\operatorname{Hom}_M(E_\mu, E_{\mu_\delta}) = \bigoplus_{l \in \mathbb{N}/2} E(-l^2)$$

for the decomposition of $\mathrm{Hom}_M(E_\mu, E_{\mu_\delta})$ into generalized eigenspaces for the action of $Z^2_{\alpha_j}$ (by $T \mapsto T \circ d\mu(Z_{\alpha_j})^2$). The factor $l_\mu(s_{\alpha_j}, \nu_{j-1})$ acts on $E(-l^2)$ by

$$T \mapsto c_l\, \mu_\delta(\sigma_{\alpha_j}) T \mu(\sigma_{\alpha_j}^{-1}),$$

where $c_l = c_l(\alpha_j, \nu_{j-1})$ are the scalars introduced in the previous section.

The operator $l_\mu(w_0, \nu)$ carries all the signature information for the Hermitian form $\langle , \rangle$ on the μ-isotypic subspace of the principal series representation $I_P(\delta_k, \nu)$, for every M-type δ_k occurring in μ_δ which is fixed by w_0.

We will compute the operator $l_\mu(w_0, \nu)$ for a particular K-type μ of $Mp(4)$, and show that it is not positive semidefinite for one choice of δ. This will prove that the nondegenerate G-invariant Hermitian form on $\overline{X}(\delta, \nu)$ is indefinite, hence the representation is nonunitary.

5.3. The Groups G^α for $Mp(4)$. Let

$$\Delta = \Delta(\mathfrak{g}, \mathfrak{a}) = \{\pm e_1 \pm e_2,\ \pm 2e_1,\ \pm 2e_2\} \tag{5.37}$$

be the set of roots of $\mathfrak{g}$ with respect to $\mathfrak{a}$. Recall that for each root $\alpha \in \Delta$ we can define a map

$$\phi_\alpha \colon \mathfrak{sl}(2, \mathbb{R}) \to \mathfrak{g}_0 \tag{5.38}$$

as in (4.3.5) of [**16**]. The image is a subalgebra of $\mathfrak{g}_0$ isomorphic to $\mathfrak{sl}(2, \mathbb{R})$. Let G_α be the corresponding connected subgroup of $Mp(4)$. Then G_α is isomorphic to either $SL(2, \mathbb{R})$ or $Mp(2)$ (since G_α is the identity component of the inverse image under pr of the corresponding subgroup of $Sp(4, \mathbb{R})$). The map ϕ_α exponentiates to a map

$$\Phi_\alpha \colon Mp(2) \to G \tag{5.39}$$

with image G_α. We define

$$Z_\alpha = \phi_\alpha \begin{pmatrix} 0 & 1 \\ -1 & 0 \end{pmatrix} \in \mathfrak{k}_0 \tag{5.40}$$

$$\sigma_\alpha = \exp\left(\frac{\pi}{2} Z_\alpha\right) \in K \quad \text{and} \tag{5.41}$$

$$m_\alpha = \exp(\pi Z_\alpha) = \sigma_\alpha^2 \in M. \tag{5.42}$$

Note that σ_α is a representative in $Mp(4)$ for the root reflection s_α; its projection is a representative in $Sp(4)$ for s_α:

$$pr(\sigma_\alpha) = \Phi_{\alpha,L} \begin{pmatrix} 0 & 1 \\ -1 & 0 \end{pmatrix},$$

with $\Phi_{\alpha,L}$ the corresponding map for the linear groups.

We now write down the maps ϕ_α and the elements Z_α, σ_α and m_α for $Mp(4)$. For $\alpha = 2e_1$,

$$\phi_{2e_1}\begin{pmatrix} u & v \\ w & -u \end{pmatrix} = \left(\begin{array}{cc|cc} u & 0 & v & 0 \\ 0 & 0 & 0 & 0 \\ \hline w & 0 & -u & 0 \\ 0 & 0 & 0 & 0 \end{array}\right) \tag{5.43}$$

$$Z_{2e_1} = \left(\begin{array}{cc|cc} 0 & 0 & 1 & 0 \\ 0 & 0 & 0 & 0 \\ \hline -1 & 0 & 0 & 0 \\ 0 & 0 & 0 & 0 \end{array}\right) = \left[\begin{pmatrix} i & 0 \\ 0 & 0 \end{pmatrix}, \frac{i}{2}\right] \tag{5.44}$$

$$\sigma_{2e_1} = \exp\left[\begin{pmatrix} \frac{\pi}{2}i & 0 \\ 0 & 0 \end{pmatrix}, \frac{\pi}{4}i\right] = \left[\begin{pmatrix} i & 0 \\ 0 & 1 \end{pmatrix}, e^{\frac{\pi}{4}i}\right] \tag{5.45}$$

$$m_{2e_1} = \left[\begin{pmatrix} -1 & 0 \\ 0 & 1 \end{pmatrix}, i\right] = a. \tag{5.46}$$

For $\alpha = 2e_2$,

$$\phi_{2e_2}\begin{pmatrix} u & v \\ w & -u \end{pmatrix} = \left(\begin{array}{cc|cc} 0 & 0 & 0 & 0 \\ 0 & u & 0 & v \\ \hline 0 & 0 & 0 & 0 \\ 0 & w & 0 & -u \end{array}\right) \tag{5.47}$$

$$Z_{2e_2} = \left(\begin{array}{cc|cc} 0 & 0 & 0 & 0 \\ 0 & 0 & 0 & 1 \\ \hline 0 & 0 & 0 & 0 \\ 0 & -1 & 0 & 0 \end{array}\right) = \left[\begin{pmatrix} 0 & 0 \\ 0 & i \end{pmatrix}, \frac{i}{2}\right] \tag{5.48}$$

$$\sigma_{2e_2} = \exp\left[\begin{pmatrix} 0 & 0 \\ 0 & \frac{\pi}{2}i \end{pmatrix}, \frac{\pi}{4}i\right] = \left[\begin{pmatrix} 1 & 0 \\ 0 & i \end{pmatrix}, e^{\frac{\pi}{4}i}\right] \tag{5.49}$$

$$m_{2e_2} = \left[\begin{pmatrix} 1 & 0 \\ 0 & -1 \end{pmatrix}, i\right] = b. \tag{5.50}$$

For $\alpha = e_1 - e_2$,

$$\phi_{e_1-e_2}\begin{pmatrix} u & v \\ w & -u \end{pmatrix} = \left(\begin{array}{cc|cc} u & v & 0 & 0 \\ w & -u & 0 & 0 \\ \hline 0 & 0 & -u & -w \\ 0 & 0 & -v & u \end{array}\right) \tag{5.51}$$

$$Z_{e_1-e_2} = \left(\begin{array}{cc|cc} 0 & 1 & 0 & 0 \\ -1 & 0 & 0 & 0 \\ \hline 0 & 0 & 0 & 1 \\ 0 & 0 & -1 & 0 \end{array}\right) = \left[\begin{pmatrix} 0 & 1 \\ -1 & 0 \end{pmatrix}, 0\right] \tag{5.52}$$

$$\sigma_{e_1-e_2} = \exp\left[\begin{pmatrix} 0 & \frac{\pi}{2} \\ -\frac{\pi}{2} & 0 \end{pmatrix}, 0\right] = \left[\begin{pmatrix} 0 & 1 \\ -1 & 0 \end{pmatrix}, 1\right] \tag{5.53}$$

$$m_{e_1-e_2} = \left[\begin{pmatrix} -1 & 0 \\ 0 & -1 \end{pmatrix}, 1\right] = y. \tag{5.54}$$

Finally, for $\alpha = e_1 + e_2$,

$$\phi_{e_1+e_2}\begin{pmatrix} u & v \\ w & -u \end{pmatrix} = \left(\begin{array}{cc|cc} u & 0 & 0 & v \\ 0 & u & v & 0 \\ \hline 0 & w & -u & 0 \\ w & 0 & 0 & -u \end{array}\right) \tag{5.55}$$

$$Z_{e_1+e_2} = \left(\begin{array}{cc|cc} 0 & 0 & 0 & 1 \\ 0 & 0 & 1 & 0 \\ \hline 0 & -1 & 0 & 0 \\ -1 & 0 & 0 & 0 \end{array}\right) = \left[\begin{pmatrix} 0 & i \\ i & 0 \end{pmatrix}, 0\right] \tag{5.56}$$

$$\sigma_{e_1+e_2} = \exp\left[\begin{pmatrix} 0 & \frac{\pi}{2}i \\ \frac{\pi}{2}i & 0 \end{pmatrix}, 0\right] = \left[\begin{pmatrix} 0 & i \\ i & 0 \end{pmatrix}, 1\right] \tag{5.57}$$

$$m_{e_1+e_2} = \left[\begin{pmatrix} -1 & 0 \\ 0 & -1 \end{pmatrix}, 1\right] = y. \tag{5.58}$$

For any root α, there is a surjective map

$$pr_\alpha \colon G_\alpha \to SL(2, \mathbb{R}), \tag{5.59}$$

which is either injective or two-to-one, depending on whether G_α is linear or not. The element m_α belongs to the inverse image of $m = diag(-1, -1)$ under this map, so $pr_\alpha(m_\alpha^2) = 1$. Note that m_α has order two if pr_α is injective, and has order four otherwise. Therefore, the order of m_α determines whether G_α is isomorphic to $SL(2)$ or $Mp(2)$. We find:

$$G_{2e_1} \simeq G_{2e_2} \simeq Mp(2) \tag{5.60}$$

and

$$G_{e_1-e_2} \simeq G_{e_1+e_2} \simeq SL(2). \tag{5.61}$$

We choose

$$\Delta^+(\mathfrak{g}, \mathfrak{a}) = \{2e_1, 2e_2, e_1 \pm e_2\} \tag{5.62}$$

so that the simple roots are

$$\alpha = e_1 - e_2 \quad \text{and} \quad \beta = 2e_2. \tag{5.63}$$

Then

$$w_0 = s_\alpha s_\beta s_\alpha s_\beta \tag{5.64}$$

is a minimal decomposition of the long Weyl group element w_0 as a product of simple reflections.

5.4. The Operator on the K-Type $\mu = \left(\frac{3}{2}, \frac{1}{2}\right)$. We will compute the operator $l(w_0, \nu)$ for the K-type $\mu = \left(\frac{3}{2}, \frac{1}{2}\right)$. First, we need to understand the structure of μ_δ and μ, and their restriction to M.

5.4.1. *The K-Type μ_δ.* The K-type μ_δ has highest weight

$$\left(\frac{1}{2}, -\frac{1}{2}\right) = (1, 0) + \left(-\frac{1}{2}, -\frac{1}{2}\right)$$

and is isomorphic to $\mathbb{C}^2 \otimes \det^{-\frac{1}{2}}$. Here $\mathbb{C}^2$ denotes the standard representation of $U(2)$. So an element $(g, z) \in \widetilde{U}(2)$ acts on a vector $v \in E_{\mu_\delta} \simeq \mathbb{C}^2$ by

$$\mu_\delta(g, z)\, v := z^{-1} g v.$$

We choose the basis $\{b_1, b_2\}$ of E_{μ_δ}, with $b_1 = \begin{bmatrix} 1 \\ 0 \end{bmatrix}$ and $b_2 = \begin{bmatrix} 0 \\ 1 \end{bmatrix}$. Then

$$\mu_\delta(a) = \mu_\delta\left[\begin{pmatrix} -1 & 0 \\ 0 & 1 \end{pmatrix}, i\right] = \begin{pmatrix} i & 0 \\ 0 & -i \end{pmatrix}, \text{ and} \tag{5.65}$$

$$\mu_\delta(b) = \mu_\delta\left[\begin{pmatrix} 1 & 0 \\ 0 & -1 \end{pmatrix}, i\right] = \begin{pmatrix} -i & 0 \\ 0 & i \end{pmatrix}. \tag{5.66}$$

Because the elements a and b generate M, the subspaces $\langle b_1 \rangle$ and $\langle b_2 \rangle$ are clearly stable under the action of M. It is easy to see that

$$\langle b_1 \rangle \simeq \delta_1 \quad \text{and} \quad \langle b_2 \rangle \simeq \delta_2.$$

The matrices of $\mu_\delta(\sigma_\alpha)$ and $\mu_\delta(\sigma_\beta)$ with respect to the basis $\{b_1, b_2\}$ are as follows.

$$\mu_\delta(\sigma_\alpha) = \mu_\delta\left[\begin{pmatrix} 0 & 1 \\ -1 & 0 \end{pmatrix}, 1\right] = \begin{pmatrix} 0 & 1 \\ -1 & 0 \end{pmatrix} \tag{5.67}$$

$$\mu_\delta(\sigma_\beta) = \mu_\delta\left[\begin{pmatrix} 1 & 0 \\ 0 & i \end{pmatrix}, e^{\frac{\pi i}{4}}\right] = \begin{pmatrix} e^{-\frac{\pi i}{4}} & 0 \\ 0 & e^{\frac{\pi i}{4}} \end{pmatrix}. \tag{5.68}$$

5.4.2. *The K-type μ.* The K-type μ has highest weight

$$\left(\frac{3}{2}, \frac{1}{2}\right) = (1, 0) + \left(\frac{1}{2}, \frac{1}{2}\right)$$

and is isomorphic to $\mathbb{C}^2 \otimes \det^{\frac{1}{2}}$. So an element $(g, z) \in \widetilde{U}(2)$ acts on a vector $v \in E_\mu \simeq \mathbb{C}^2$ by

$$\mu(g, z)\, v := z\, gv.$$

The differentiated action is

$$\mathrm{d}\mu(Z, z)\, v = (Z + zI)\, v,$$

with I the identity matrix.

We choose the basis $\{f_1, f_2\}$ of E_μ, with $f_1 = \begin{bmatrix} 1 \\ 0 \end{bmatrix}$ and $f_2 = \begin{bmatrix} 0 \\ 1 \end{bmatrix}$. Then

$$\mu(a) = \mu\left[\begin{pmatrix} -1 & 0 \\ 0 & 1 \end{pmatrix}, i\right] = \begin{pmatrix} -i & 0 \\ 0 & i \end{pmatrix}, \text{ and} \tag{5.69}$$

$$\mu(b) = \mu\left[\begin{pmatrix} 1 & 0 \\ 0 & -1 \end{pmatrix}, i\right] = \begin{pmatrix} i & 0 \\ 0 & -i \end{pmatrix}. \tag{5.70}$$

The subspaces $\langle f_1 \rangle$ and $\langle f_2 \rangle$ are M-invariant, and

$$\langle f_1 \rangle \simeq \delta_2 \quad \text{and} \quad \langle f_2 \rangle \simeq \delta_1.$$

We give the matrices of $\mu(\sigma_\alpha^{-1})$, $\mathrm{d}\mu(Z_\alpha)$, $\mathrm{d}\mu(Z_\alpha)^2$ and $\mu\left(\sigma_\beta^{-1}\right)$, $\mathrm{d}\mu(Z_\beta)$ and $\mathrm{d}\mu(Z_\beta)^2$ with respect to the basis $\{f_1, f_2\}$.

$$\mu(\sigma_\alpha^{-1}) = \mu\left[\begin{pmatrix} 0 & -1 \\ 1 & 0 \end{pmatrix}, 1\right] = \begin{pmatrix} 0 & -1 \\ 1 & 0 \end{pmatrix} \tag{5.71}$$

$$\mathrm{d}\mu(Z_\alpha) = \mathrm{d}\mu\left[\begin{pmatrix} 0 & 1 \\ -1 & 0 \end{pmatrix}, 0\right] = \begin{pmatrix} 0 & 1 \\ -1 & 0 \end{pmatrix} \tag{5.72}$$

$$\mathrm{d}\mu(Z_\alpha)^2 = \begin{pmatrix} -1 & 0 \\ 0 & -1 \end{pmatrix} \tag{5.73}$$

and

$$\mu\left(\sigma_\beta^{-1}\right) = \mu\left[\begin{pmatrix} 1 & 0 \\ 0 & -i \end{pmatrix}, e^{-\frac{\pi i}{4}}\right] = \begin{pmatrix} e^{-\frac{\pi i}{4}} & 0 \\ 0 & e^{-\frac{3\pi i}{4}} \end{pmatrix} \tag{5.74}$$

$$\mathrm{d}\mu\left(Z_\beta\right) = \mathrm{d}\mu\left[\begin{pmatrix} 0 & 0 \\ 0 & i \end{pmatrix}, \frac{i}{2}\right] = \begin{pmatrix} \frac{i}{2} & 0 \\ o & \frac{3i}{2} \end{pmatrix} \tag{5.75}$$

$$\mathrm{d}\mu\left(Z_\beta\right)^2 = \begin{pmatrix} -\frac{1}{4} & 0 \\ 0 & -\frac{9}{4} \end{pmatrix}. \tag{5.76}$$

5.4.3. *The Operator $l_\mu(w_0, \nu)$ on $\mathrm{Hom}_M(E_\mu, E_{\mu_\delta})$.* We choose the basis

$$\begin{array}{ll} T_1(f_1) = 0 & T_2(f_1) = b_2 \\ T_1(f_2) = b_1 & T_2(f_2) = 0 \end{array}$$

of the space $E := \mathrm{Hom}_M\left(V^\mu, V^{\mu_\delta}\right)$.
Note that $T_1 \in \mathrm{Hom}_M\left(V^\mu, E_{\mu_\delta}(\delta_1)\right)$, and $T_2 \in \mathrm{Hom}_M\left(V^\mu, E_{\mu_\delta}(\delta_2)\right)$.

We compute the operator

$$l_\mu(s_\gamma, \lambda) \colon E \to E$$

associated to each simple reflection γ. Recall the recipe:

- Decompose E into eigenspaces for the action of $\mathrm{d}\mu(Z_\gamma)^2$: $E = \bigoplus_{l \in \mathbb{N}/2} E(-l^2)$, with
$$E(-l^2) := \{T \in E \colon T \circ \mathrm{d}\mu(Z_\gamma)^2 = (-l^2)T\}$$
- The operator $l_\mu(s_\gamma, \lambda)$ acts on an element $T \in E(-l^2)$ by
$$l_\mu(s_\gamma, \lambda)T = c_l(\gamma, \lambda)\, \mu_\delta(\sigma_\gamma) T \mu(\sigma_\gamma)^{-1},$$
with
$$c_l(\gamma, \lambda) = \begin{cases} 1 & \text{if } l = 0, 1, \text{or } \frac{1}{2} \\ -\frac{\frac{1}{2} - \langle \lambda, \gamma^\vee \rangle}{\frac{1}{2} + \langle \lambda, \gamma^\vee \rangle} & \text{if } l = \frac{3}{2} \\ -\frac{1 - \langle \lambda, \gamma^\vee \rangle}{1 + \langle \lambda, \gamma^\vee \rangle} & \text{if } l = 2 \end{cases} \tag{5.77}$$
(for our purpose we do not need the constants c_l for other values of l).
For brevity of notation, we write: $\mu_\delta(\sigma_\gamma) T \mu(\sigma_\gamma)^{-1} := \psi_\mu(s_\gamma) T$.

We start by computing the operator $l_\mu(s_\alpha, \lambda)$ (for $\alpha = e_1 - e_2$).
Recall that $\mathrm{d}\mu(Z_\alpha^2) = -I$, so for all $T \in E$ and all $v \in E_\mu$, we have

$$T\left(\mathrm{d}\mu(Z_\alpha)^2 v\right) = -T(v).$$

Hence $E = E_{-1}$ and $l_\mu(s_\alpha, \lambda) \equiv \psi_\mu\left(s_\alpha\right)$.
For all $v = a_1 f_1 + a_2 f_2 \in E_\mu$, we compute:

$$\psi_\mu\left(s_\alpha\right) T_1\left(v\right) = \mu_\delta(\sigma_\alpha) T_1\left(\mu\left(\sigma_\alpha^{-1}\right) v\right) = \begin{pmatrix} 0 & 1 \\ -1 & 0 \end{pmatrix} T_1\left(\begin{pmatrix} 0 & -1 \\ 1 & 0 \end{pmatrix}\begin{pmatrix} a_1 \\ a_2 \end{pmatrix}\right) =$$
$$= \begin{pmatrix} 0 & 1 \\ -1 & 0 \end{pmatrix} T_1\begin{pmatrix} -a_2 \\ a_1 \end{pmatrix} = \begin{pmatrix} 0 & 1 \\ -1 & 0 \end{pmatrix}\begin{pmatrix} a_1 \\ 0 \end{pmatrix} = \begin{pmatrix} 0 \\ -a_1 \end{pmatrix} = -T_2(v).$$

Similarly,

$$\psi_\mu\left(s_\alpha\right) T_2\left(v\right) = \begin{pmatrix} 0 & 1 \\ -1 & 0 \end{pmatrix} T_2\left(\begin{pmatrix} 0 & -1 \\ 1 & 0 \end{pmatrix}\begin{pmatrix} a_1 \\ a_2 \end{pmatrix}\right) =$$
$$= \begin{pmatrix} 0 & 1 \\ -1 & 0 \end{pmatrix} T_2\begin{pmatrix} -a_2 \\ a_1 \end{pmatrix} = \begin{pmatrix} 0 & 1 \\ -1 & 0 \end{pmatrix}\begin{pmatrix} 0 \\ -a_2 \end{pmatrix} = \begin{pmatrix} -a_2 \\ 0 \end{pmatrix} = -T_1(v).$$

Then

$$l_\mu(s_\alpha, \lambda) = \begin{pmatrix} 0 & -1 \\ -1 & 0 \end{pmatrix}. \tag{5.78}$$

Now we do the corresponding computations for $\beta = 2e_2$.
For all $j = 1, 2$ and all $v \in E_\mu$, we have:

$$T_j\left(\mathrm{d}\mu(Z_\beta)^2 v\right) = T_j\left(\begin{pmatrix} -\frac{1}{4} & 0 \\ 0 & -\frac{9}{4} \end{pmatrix}\begin{pmatrix} a_1 \\ a_2 \end{pmatrix}\right) = T_j\begin{pmatrix} -\frac{1}{4}a_1 \\ -\frac{9}{4}a_2 \end{pmatrix} = \begin{cases} -\frac{9}{4}T_j & \text{if } j = 1 \\ -\frac{1}{4}T_j & \text{if } j = 2 \end{cases}$$

so $E \simeq E\left(-\frac{1}{4}\right) \oplus E\left(-\frac{9}{4}\right)$, with

$$E\left(-\frac{1}{4}\right) = \langle T_2 \rangle \quad \text{and} \quad E\left(-\frac{9}{4}\right) = \langle T_1 \rangle.$$

It follows that

$$l_\mu(s_\beta, \lambda)T_j = \begin{cases} c_{3/2}(\beta, \lambda)\, \psi_\mu(s_\beta)T_j & \text{if } j = 1 \\ \psi_\mu(s_\beta)T_j & \text{if } j = 2. \end{cases}$$

Recall that $c_{3/2}(\beta, \lambda) = -\frac{\frac{1}{2} - \langle \lambda, \beta^\vee \rangle}{\frac{1}{2} + \langle \lambda, \beta^\vee \rangle}$. We now compute $\psi_\mu(s_\beta)T_j$, for $j = 1, 2$.

$$\begin{aligned}
\psi_\mu(s_\beta)\, T_j(v) &= \mu_\delta(\sigma_\beta) T_j\left(\mu\left(\sigma_\beta^{-1}\right) v\right) \\
&= \begin{pmatrix} e^{-\frac{\pi i}{4}} & 0 \\ 0 & e^{\frac{\pi i}{4}} \end{pmatrix} T_j\left(\begin{pmatrix} e^{-\frac{\pi i}{4}} & 0 \\ 0 & e^{-\frac{3\pi i}{4}} \end{pmatrix}\begin{pmatrix} a_1 \\ a_2 \end{pmatrix}\right) \\
&= \begin{pmatrix} e^{-\frac{\pi i}{4}} & 0 \\ 0 & e^{\frac{\pi i}{4}} \end{pmatrix} T_j\begin{pmatrix} e^{-\frac{\pi i}{4}}a_1 \\ e^{-\frac{3\pi i}{4}}a_2 \end{pmatrix} \\
&= \begin{cases} -T_j(v) & \text{if } j = 1 \\ +T_j(v) & \text{if } j = 2. \end{cases}
\end{aligned}$$

Hence we get:

$$l_\mu(s_\beta, \lambda)T_j = \begin{cases} +\frac{\frac{1}{2} - \langle \lambda, \beta^\vee \rangle}{\frac{1}{2} + \langle \lambda, \beta^\vee \rangle} T_j & \text{if } j = 1 \\ T_j & \text{if } j = 2. \end{cases}$$

Equivalently,

$$l_\mu(s_\beta, \lambda) = \begin{pmatrix} \frac{\frac{1}{2} - \langle \lambda, \beta^\vee \rangle}{\frac{1}{2} + \langle \lambda, \beta^\vee \rangle} & 0 \\ 0 & 1 \end{pmatrix}. \tag{5.79}$$

We are now ready to compute the full intertwining operator:

$$l_\mu(w_0, \nu) = l_\mu(s_\alpha, s_\beta s_\alpha s_\beta \nu)\, l_\mu(s_\beta, s_\alpha s_\beta \nu)\, l_\mu(s_\alpha, s_\beta \nu)\, l_\mu(s_\beta, \nu).$$

Write $\nu = (\nu_1, \nu_2)$, with $\nu_1 \geq \nu_2 \geq 0$. Then equations (5.78) and (5.79) give:

- $l_\mu(s_\alpha, s_\beta s_\alpha s_\beta \nu) = l_\mu(s_\alpha, s_\beta \nu) = \begin{pmatrix} 0 & -1 \\ -1 & 0 \end{pmatrix}$

- $l_{\mu}(s_{\beta},\nu)=\begin{pmatrix}\frac{\frac{1}{2}-\langle\nu,\beta^{\vee}\rangle}{\frac{1}{2}+\langle\nu,\beta^{\vee}\rangle} & 0\\ 0 & 1\end{pmatrix}=\begin{pmatrix}\frac{\frac{1}{2}-\nu_2}{\frac{1}{2}+\nu_2} & 0\\ 0 & 1\end{pmatrix}$

- $l_{\mu}(s_{\beta},s_{\alpha}s_{\beta}\nu)=\begin{pmatrix}\frac{\frac{1}{2}-\langle s_{\alpha}s_{\beta}\nu,\beta^{\vee}\rangle}{\frac{1}{2}+\langle s_{\alpha}s_{\beta}\nu,\beta^{\vee}\rangle} & 0\\ 0 & 1\end{pmatrix}=\begin{pmatrix}\frac{\frac{1}{2}-\nu_1}{\frac{1}{2}+\nu_1} & 0\\ 0 & 1\end{pmatrix}$

and we obtain:

$$l_{\mu}(w_0,\nu)=\begin{pmatrix}\frac{\frac{1}{2}-\nu_2}{\frac{1}{2}+\nu_2} & 0\\ 0 & \frac{\frac{1}{2}-\nu_1}{\frac{1}{2}+\nu_1}\end{pmatrix}. \tag{5.80}$$

For all $i=1,2$, let $\bar{X}(\delta_i,\nu)$ be the irreducible constituent of the principal series $I_P(\delta_i,\nu)$ containing the lowest K-type $\mu_\delta=\left(\frac{1}{2},-\frac{1}{2}\right)$. The operator $l_\mu(w_0,\nu)$ carries the signature information on the K-type $\mu=\left(\frac{3}{2},\frac{1}{2}\right)$ for *both* $\bar{X}(\delta_1,\nu)$ and $\bar{X}(\delta_2,\nu)$, and can be interpreted as the direct sum of the operators

$$l_\mu(w_0,\delta_1,\nu)\colon \operatorname{Hom}_M(\nu,\delta_1)\to\operatorname{Hom}_M(\nu,\delta_1)$$

and

$$l_\mu(w_0,\delta_2,\nu)\colon \operatorname{Hom}_M(\nu,\delta_2)\to\operatorname{Hom}_M(\nu,\delta_2).$$

Recall that $l_{\mu_\delta}(w_0,\delta_i,\nu)=1$ for both $i=1,2$, because of our normalization. We can sometimes use the operator $l_\mu(w_0,\nu)$ to detect the nonunitarity of $\bar{X}(\delta_i,\nu)$:

REMARK 5.1. If the ith-entry of $l_\mu(w_0,\nu)$ is negative, then the nondegenerate Hermitian form on $\bar{X}(\delta_i,\nu)$ is indefinite, and the representation is nonunitary. If the ith-entry of $l_\mu(w_0,\nu)$ is zero, then the K-type μ does not appear in $\bar{X}(\delta_i,\nu)$. If the ith-entry of $l_\mu(w_0,\nu)$ is positive, then we cannot draw any conclusion regarding the unitarity of $\bar{X}(\delta_i,\nu)$.

We are interested in the Mystery representation, that has infinitesimal character $\nu=\left(\frac{3}{2},\frac{1}{2}\right)$. In this case

$$l_\mu(w_0,\nu)=\begin{pmatrix}0 & 0\\ 0 & -\frac{1}{2}\end{pmatrix}$$

hence the previous remark implies that $\overline{X}(\delta_2,\nu)$ is nonunitary.

THEOREM 5.2. *The Mystery representation of $Mp(4)$ is nonunitary.*

Consequently, we obtain the following result:

THEOREM 5.3. *Let X be an irreducible admissible representation of $Mp(4)$. Then X is unitary and ω-regular if and only if X is either an $A_{\mathfrak{q}}(\Omega)$ or a Meta-$A_{\mathfrak{q}}(\lambda)$ representation.*

6. Appendix

6.1. Lie Groups and Representations. Consider figures which have continuous sets of symmetries, such as the circle or the sphere. For any small number x, the rotation $k_x=\begin{pmatrix}\cos x & \sin x\\ -\sin x & \cos x\end{pmatrix}$ moves the points on the circle by an angle x. We get a one-dimensional set S^1 of rotations that leave the circle looking the same. Similarly, we can draw an imaginary axis of rotation from any point on the sphere to its center. That gives us two dimensions of choices for the axis of rotation plus one dimension for the angle of rotation around the chosen axis; so the rotations of a sphere form a three-dimensional set S^2.

A Lie Group is a continuous group of symmetries. More precisely, it is a C^∞ manifold with a smooth group structure. The sets S^1 and S^2 of rotations of the circle and the sphere are examples of one- and three-dimensional Lie groups. If V is an n-dimensional vector space over a field $\mathbb{F}$, then the set $Aut\,(V) \simeq GL\,(n, \mathbb{F})$ is a Lie group of dimension n^2.

The group S^1 acts on a one-dimensional space by scalars: $k_t u = \lambda\,(t)\,u$, where $\lambda\,(t)$ is a complex number. S^1 also acts on the plane by multiplication:

$$k_t \begin{pmatrix} u \\ v \end{pmatrix} = \begin{pmatrix} \cos t & \sin t \\ -\sin t & \cos t \end{pmatrix} \begin{pmatrix} u \\ v \end{pmatrix} = \begin{pmatrix} u' \\ v' \end{pmatrix}, \forall\, u, v \in \mathbb{C}$$

and on functions (e.g. L^2 functions) on the unit circle by left translation:

$$k_t f \left(\begin{pmatrix} u \\ v \end{pmatrix} \right) = f \left(k_t^{-1} \begin{pmatrix} u \\ v \end{pmatrix} \right), \forall\, u, v \in \mathbb{R} \ s.t. \ u^2 + v^2 = 1.$$

The latter is an infinite dimensional vector space.

Representation Theory studies all possible ways in which a Lie group acts on vector spaces. We call all these possibilities the representations of the group. More precisely, a representation of a group G is a continuous homomorphism

$$\phi \colon G \to Aut\,(V) \tag{6.1}$$

for some vector space V. In case all group elements act by the identity operator, we call the representation trivial.

The representations of the rotation group S^1 are related to Fourier series: all the basic (irreducible) representations are one-dimensional, of periodic type. The action is

$$\phi_n\,(k_t)\,v = e^{int} v, \ \forall\, v \in \mathbb{C}.$$

Applications to science abound. Examples include areas where we use infinite series of the form $f\,(t) = \sum_{n \in \mathbb{Z}} a_n e^{int}$.

A representation of G on V is called irreducible if the vector space V does not have proper, closed subspaces which are themselves left invariant by the group. Irreducible representations are the building blocks of bigger representations.

REMARK 6.1. Irreducible representations of compact groups are finite dimensional. Irreducible representations of abelian groups are one-dimensional.

EXAMPLE 6.2. All irreducible representations of the diagonal group

$$A = \left\{ \begin{pmatrix} a & 0 \\ 0 & a^{-1} \end{pmatrix} : a > 0,\, a \in \mathbb{R} \right\} \tag{6.2}$$

are given by

$$\phi_r \begin{pmatrix} a & 0 \\ 0 & a^{-1} \end{pmatrix} v = a^r v$$

for $r \in \mathbb{C}$.

DEFINITION 6.3. A representation $\phi \colon G \to Aut\,(V)$ of G on a vector space V is unitary if V is a Hilbert space and all operators $\phi\,(g)$ are unitary operators on V (i.e. preserve the inner product on V).

EXAMPLE 6.4. Square integrable functions on the circle $L^2\,(S^1)$ form a unitary representation of the rotation group S^1.

EXAMPLE 6.5. All irreducible (hence finite dimensional) representations of a compact group are unitary.

EXAMPLE 6.6. A representation ϕ_r of the diagonal group A is unitary if r is purely imaginary.

6.2. The group $SL(2,\mathbf{R})$. Consider the group $G = SL(2,\mathbb{R})$ of 2×2 real matrices with determinant one. In order to construct a large family of representations of $SL(2,\mathbb{R})$, we consider the following subgroups:

$$M = \{\pm I\} \tag{6.3}$$

$$A = \left\{\begin{pmatrix} a & 0 \\ 0 & a^{-1} \end{pmatrix} : a > 0\right\} \tag{6.4}$$

$$N = \left\{\begin{pmatrix} 1 & x \\ 0 & 1 \end{pmatrix} : x \in \mathbb{R}\right\}$$

$$K = \left\{\begin{pmatrix} \cos t & \sin t \\ -\sin t & \cos t \end{pmatrix} : t \in \mathbb{R}\right\} = \left\{\exp\begin{pmatrix} 0 & t \\ -t & 0 \end{pmatrix} : t \in \mathbb{R}\right\} \equiv S^1$$

REMARK 6.7. The irreducible representations of M are $\{\delta_+, \delta_- : \delta_\pm(-I) = \pm 1\}$. Those of N are also one-dimensional. In our case, N will act trivially, i. e., by fixing all vectors.

6.2.1. *The Lie algebra of $SL(2,\mathbb{R})$.* The set

$$\mathfrak{g} = \left\{\begin{pmatrix} a & b \\ c & -a \end{pmatrix} : a, b, c \in \mathbb{C}\right\} = \mathfrak{sl}(2,\mathbb{C}) \tag{6.5}$$

is a linear vector space, called the complexified Lie algebra of $SL(2,\mathbb{R})$. The real Lie algebra $\mathfrak{g}_0$, with real entries, is the tangent space at the identity of the Lie group $SL(2,\mathbb{R})$. We will use similar notation for other Lie algebras, e.g. $\mathfrak{k}$ and $\mathfrak{k}_0$ are the complexified and the real Lie algebra of K.
The space $\mathfrak{g}$ has a (bilinear) bracket operation $[x, y] = xy - yx$ and basis elements

$$H = -i\begin{pmatrix} 0 & 1 \\ -1 & 0 \end{pmatrix}, X = \frac{1}{2}\begin{pmatrix} 1 & i \\ i & -1 \end{pmatrix}, Y = \frac{1}{2}\begin{pmatrix} 1 & -i \\ -i & -1 \end{pmatrix}.$$

They satisfy:

$$\begin{array}{ll} [x,x] = 0 & [H,X] = 2X \\ [x,[y,z]] = [[x,y],z] + [y,[x,z]] & [H,Y] = -2Y \\ & [X,Y] = H. \end{array}$$

Note that if we define $\mathfrak{g}_\lambda = \{Z \in \mathfrak{g} \colon [H, Z] = \lambda Z\}$, then $\mathfrak{g} = \mathfrak{g}_0 + \mathfrak{g}_2 + \mathfrak{g}_{-2}$.

DEFINITION 6.8. A Lie algebra representation is a homomorphism $\pi \colon \mathfrak{g} \to End(W)$ for some vector space W, preserving the bracket operations:

$$\pi\left([x,y]_{\mathfrak{g}}\right) = [\pi(x), \pi(y)]_{End(W)} = \pi(x)\pi(y) - \pi(y)\pi(x) \quad \forall x, y \in \mathfrak{g}.$$

If W is finite dimensional, then $\pi(x) : W \to W$ is a matrix for all $x \in \mathfrak{g}$.

6.2.2. *Harish-Chandra modules of* $G = SL(2,\mathbb{R})$. If $\phi\colon G \to Aut\,(V)$ is a representation of G, we can restrict the action of the group on V to any subgroup. In particular, we can restrict the representation to the maximal compact group $K = S^1$, since we know a bit more about its representations. This way we hope to be able to say something about the representation of G.

A representation ϕ of G is called admissible if, when restricted the maximal compact group K, each irreducible representation μ of K occurs with finite multiplicity. We call the sum of those copies of μ the isotypic component of μ. If ϕ is admissible, then we can find a dense subspace $W \subset V$ where ϕ is differentiable. The action of $\mathfrak{g}$ on W is defined by

$$\pi\,(Z)\,w = \frac{d}{dt}\phi\,(\exp tZ)\,|_{t=0}w \tag{6.6}$$

for all Z in $\mathfrak{g}$ and w in W.
Suppose that (ϕ, V) is an admissible representation of G. The restriction of ϕ to the compact subgroup K is a Hilbert space direct sum of isotypic components of irreducible representations of K:

$$V = \widehat{\bigoplus_{n\in\mathbb{Z}\colon\, V(n)\neq 0}} V(n)$$

(recall that the irreducible representations $\{\phi_n\}$ of $K = S^1$ are parameterized by integers). Set

$$W = \bigoplus_{n\in\mathbb{Z}\colon\, V(n)\neq 0} V(n)$$

(the algebraic direct sum of the $V(n)$'s). Then W is a dense subspace of V and ϕ is differentiable on W. Hence we have an action π of $\mathfrak{g}$ on W, as in (6.6).

REMARK 6.9. For all $w \in V(n) \subset W$

$$\begin{aligned}\pi\,(iH)\,w &= \frac{d}{dt}\phi_n\left(\exp t\begin{pmatrix} 0 & 1 \\ -1 & 0\end{pmatrix}\right)_{t=0} w \\ &= \frac{d}{dt}\left(e^{int}\right)_{t=0} w = inw.\end{aligned}$$

Hence we can identify $V(n)$ with the eigenspace for H of eigenvalue n.

The representation W is called a $(\mathfrak{g},K)$ module, or a Harish-Chandra module, of G. Let H, X, and Y be as above.

PROPOSITION 11. For any Harish-Chandra module W of $G = SL(2,\mathbb{R})$, the action of $\mathfrak{g}$ on W satisfies:

- $H \cdot V(n) \subseteq V(n)$ diagonalizable operator
- $X \cdot V(n) \subseteq V(n+2)$ "raising" operator
- $Y \cdot V(n) \subseteq V(n-2)$ "lowering" operator.

DEFINITION 6.10. Let W be a Harish-Chandra module, and let

$$n_0 := \min_{n\in\mathbb{Z}} \{|n|\colon V(n) \neq 0\}\,.$$

We say that $V(n_0)$ (or $V(-n_0)$) is the lowest K-type in W. It is the lowest irreducible representation of K occurring in $W|_K$.

It turns out that we can parameterize irreducible $(\mathfrak{g},K)$ modules by their lowest K-types. First assume $n_0 \geq 2$. If W is irreducible, then either $V(-n_0) = 0$ (and $V(n_0) \neq 0$) or $V(n_0) = 0$ (and $V(-n_0) \neq 0$). In the first case, W is generated using X from the representation $V(n_0)$ of K:

$$\begin{aligned} W &\simeq V(n_0) + V(n_0+2) + V(n_0+4) + \ldots \\ &= D(n_0). \end{aligned}$$

Here $D(n_0)$ is the discrete series representation with lowest K-type n_0. Similarly, if $V(n_0) = 0$, then we can generate W from $V(-n_0)$ using Y:

$$\begin{aligned} W &\simeq V(-n_0) + V(-n_0-2) + V(-n_0-4) + \ldots \\ &= D(-n_0). \end{aligned}$$

Now consider the case $n_0 = 0$ or 1. Any representation of G with lowest K-type 0 or 1 can be constructed from a representation of a parabolic subgroup $P = MAN$, as follows. Define

$$I_P(\delta,\nu) = \left\{ f\colon G \to V_{\delta,\nu}\colon f \in L^2(K),\ f(gman) = \delta\left(m^{-1}\right) e^{-(\nu-1)\log a} f(g) \right\}$$

where $V_{\delta,\nu}$ is the one-dimensional representation $\delta \otimes \nu \otimes \mathit{trivial}$ of P (with $\delta = \delta_\pm \in \hat{M}$, $\nu \in \mathbb{C} \simeq \hat{A}$). The group G acts on $I_P(\delta,\nu)$ by left translation:

$$\phi(x)f(g) = f(x^{-1}g) \tag{6.7}$$

for all x, $g \in G$ and $f \in I_P(\delta,\nu)$. The representation $I_P(\delta,\nu)$ is called the induced representation (or principal series) with parameters δ and ν .

REMARK 6.11. The representation $I_P(\delta,\nu)$ is not necessarily irreducible; e. g., $I_P(\delta_-,0)$ is the sum of two irreducible representations, which we call $D(\pm 1)$. Moreover, every discrete series representation is a submodule of some $I_P(\delta,\nu)$.

The properties of the principal series and the discrete representations of $G = SL(2,\mathbb{R})$ are described in Table 1.

TABLE 1

(6.8)

Representation	Irreducible	Unitary
$D(n)\,,\ \lvert n\rvert \geq 1$	*yes*	*yes*
$I_P(\delta_\pm, iv) \simeq I_P(\delta_\pm, -iv)$, $v \neq 0$, *real*	*yes*	*yes*
$I_P(\delta_-,0)$	*no*	*yes*
$I_P(\delta_+,0)$	*yes*	*yes*
$I_P(\delta_+,u)$, $0 < u < 1$	*yes*	*yes*
$I_P(\delta_\varepsilon,m)$, $m \in \mathbb{Z}$, $\varepsilon = (-1)^m$	*yes*	*no*
$I_P(\delta_\varepsilon,m)$, $m \in \mathbb{Z}$, $\varepsilon = (-1)^{m+1}$	*no*	*no*
$I_P(\delta_\pm, u+iv)$, $u, v \neq 0$	*yes*	*no*

It turns out that these two constructions can be applied to any real reductive Lie group to give an exhaustive classification of its admissible representations.

References

[1] J. Adams and D. Barbasch, *Genuine Representations of the Metaplectic Group*, Comp. Math. **113** (1998), 23-66.

[2] J. Adams, D. Barbasch, A. Paul, P. Trapa, and D. Vogan, *Unitary Shimura Correspondences for Split Real Groups*, J. Amer. Math. Soc. **20** (2007), 701-751.

[3] D. Barbasch, *Spherical Unitary Dual for Split Classical Groups,* preprint.

[4] R. Howe and E. C. Tan, *Non-Abelian Harmonic Analysis. Applications of* $SL(2,\mathbb{R})$. Universitext. New York etc., Springer-Verlag. xv, (1992).

[5] A. Knapp, *Representation Theory of Semisimple Groups. An Overview Based on Examples,* Princeton University Press, Princeton, NJ, 1986.

[6] A. Knapp and D. A. Vogan, *Cohomological Induction and Unitary Representations,* Princeton University Press, Princeton, NJ, 1995.

[7] A. Pantano, *Weyl Group Representations and Signatures of Intertwining Operators,* Princeton University thesis, 2004.

[8] R. Parthasarathy, *Criteria for the Unitarizability of Some Highest Weight Modules,* Proc. Indian Acad. Sci. **89** (1980), 1-24.

[9] A. Paul, *On the Howe Correspondence for Symplectic-Orthogonal Dual Pairs*, J. Funct. Anal. **228** (2005), 270-310.

[10] T. Przebinda, *The Oscillator Duality Correspondence for the Pair* $(O(2,2), Sp(2,\mathbb{R}))$, Mem. Am. Math. Soc. **79** (1989).

[11] S. A. Salamanca-Riba, *On the Unitary Dual of Some Classical Lie Groups,* Comp. Math. **68** (1988), 251-303.

[12] ———, *On the Unitary Dual of Real Reductive Lie Groups and the* $A_{\mathfrak{q}}(\lambda)$*-Modules: the Strongly Regular Case*, Duke Math. J. **96** (1999), no. 3, 521-546.

[13] S. A. Salamanca-Riba and D. A. Vogan Jr., *On the Classification of Unitary Representations of Reductive Lie Groups*, Ann. Math. **148** (1998), 1067-1133.

[14] B. Speh and D. A. Vogan, *Reducibility of Generalized Principal Series Representations*, Acta Math. **145** (1980), 227-299.

[15] D. A. Vogan, *The Algebraic Structure of the Representations of Semisimple Lie Groups I,* Ann. Math. **109** (1979), 1-60.

[16] ———, *Representations of Real Reductive Lie Groups*, Progress in Math. **15** (1981), Birkhäuser(Boston).

[17] ———, *Unitarizability of Certain Series of Representations*, Ann. Math. **120** (1984), 141-187.

[18] ———, *Computing the Unitary Dual,* preprint. Available at http://atlas.math.umd.edu/papers/computing.pdf.

[19] D. A. Vogan and G. Zuckerman, *Unitary Representations with Non-Zero Cohomology,* Comp. Math. **53** (1984), 51-90.

[20] C.-B. Zhu, *Representations with Scalar K-Types and Applications*, Israel J. Math. **135** (2003), 111-124.

Department of Mathematics, University of California at Irvine, Irvine, CA 92697
E-mail address: apantano@math.uci.edu

Department of Mathematics, Western Michigan University, Kalamazoo, MI 49008
E-mail address: annegret.paul@wmich.edu
URL: http://homepages.wmich.edu/~paul

Department of Mathematics, New Mexico State University, Las Cruces, NM 88003
E-mail address: ssalaman@nmsu.edu
URL: http://www.math.nmsu.edu/~ssalaman

Contemporary Mathematics
Volume **467**, 2008

A resolvent trace formula for the BTZ black hole with conical singularity

Floyd L. Williams

To the memory of Mitchell H. Taibleson

ABSTRACT. Recently we obtained an expression of the Mann-Solodukhin quantum correction to the entropy of a BTZ black hole in terms of a certain deformation of the Patterson-Selberg zeta function. As another application of this deformation, we set up a trace formula, in the presence of a conical singularity, for the resolvent of a Laplacian.

1. Introduction

In the paper [11], S. J. Patterson constructs a Selberg type zeta function $Z_\Gamma(s)$ for a discrete group Γ acting on hyperbolic n-space $\mathbb{H}^n$, where a fundamental domain for this action is of <u>infinite</u> hyperbolic volume. A derivation of a Maass-Selberg type relation for analysing Eisenstein series attached to Γ is established, of which an ingredient in the proof is a formula (Proposition 3.3 of [11]), say for a cyclic Γ, that expresses the trace of the resolvent of a Laplacian in terms of the logarithmic derivative of $Z_\Gamma(s)$. We remark on an alternate proof of this formula that employs the heat kernel of $X_\Gamma \overset{def.}{=} \Gamma\backslash\mathbb{H}^3$, especially in the case of physical interest $n = 3$ where X_Γ is a BTZ black hole, for a suitable choice of generator of Γ [1,2,4,14].

In fact for $n = 3$, we have constructed a <u>deformation</u> $\{Z_{\Gamma^{(\alpha)}}(s)\}_{0<\alpha\leq 1}$ of $Z_\Gamma(s)$ (with $Z_{\Gamma^{(1)}}(s) = Z_\Gamma(s)$) in terms of which a <u>quantum correction</u> to black hole entropy can be expressed [2,8,10,17,18]. Here the quotient $Z_{\Gamma^{(\alpha)}}\backslash\mathbb{H}^3$ for $\alpha \neq 1$ is <u>not</u> a manifold, but has a conical singularity such that by suitable differentiation at the horizon with respect to the corresponding defect angle one obtains the desired one-loop, ultra-violet correction. Moreover, using the deformation $\{Z_{\Gamma^{(\alpha)}}(s)\}_{0<\alpha\leq 1}$, we present in this paper a corresponding resolvent trace formula. That is, we obtain the Patterson formula by taking $\alpha = 1$, and more generally we obtain a trace formula in the presence of a conical singularity, at least in case $1/\alpha \in \mathbb{Z}$.

1991 *Mathematics Subject Classification.* Primary 83C57, 11M36, 47A10; Secondary 11F72, 83C05.

Key words and phrases. Patterson-Selberg zeta function,resolvent kernel,truncated heat kernel,conical singularity,BTZ black hole.

We have attempted to keep the paper purely mathematical for the most part. Only in the last section (section 5) have we specified a connection to black hole physics - the motivation for the paper.

This paper is dedicated to the memory of Professor Mitchell H. Taibleson, distinguished mathematical scholar, researcher, and teacher at Washington University, an especially valued mentor and friend to me during my student years there.

2. Resolvent of the hyperbolic Laplacian

Let $\triangle = \frac{\partial^2}{\partial x^2} + \frac{\partial^2}{\partial y^2} + \frac{\partial^2}{\partial z^2}$ denote the standard Laplacian on $\mathbb{R}^3$ with coordinates (x, y, z) and let $L = z^2\triangle - z\frac{\partial}{\partial z}$ denote the Laplacian on hyperbolic 3-space $\mathbb{H}^3 \overset{def.}{=} \{(x, y, z) \in \mathbb{R}^3 \mid z > 0\}$ with respect to the Poincaré metric

$$ds^2 = (dx^2 + dy^2 + dz^2)/z^2. \tag{2.1}$$

The corresponding hyperbolic volume element is given by

$$dv = dxdydz/z^3, \tag{2.2}$$

and the hyperbolic distance $d(p_1, p_2)$ between two points $p_j = (x_j, y_j, z_j) \in \mathbb{H}^3$, $j = 1, 2$, is given by

$$\cosh(d(p_1, p_2)) = 1 + \frac{(x_1 - x_2)^2 + (y_1 - y_2)^2 + (z_1 - z_2)^2}{2z_1 z_2}. \tag{2.3}$$

It is convenient to set

$$2\sigma(p_1, p_2) \overset{def.}{=} 1 + \cosh(d(p_1, p_2)),$$

$$u(p_1, p_2) \overset{def.}{=} \sigma(p_1, p_2) - 1 = \frac{(x_1 - x_2)^2 + (y_1 - y_2)^2 + (z_1 - z_2)^2}{4z_1 z_2}. \tag{2.4}$$

For $s \in \mathbb{C}$ we define the function ϕ_s on $\mathbb{R}^+ \overset{def.}{=} \{u \in \mathbb{R} \mid u > 0\}$ by

$$\begin{aligned} \phi_s(u) &\overset{def.}{=} \frac{1}{8\pi}(1+u)^{-s}(1+\frac{1}{u})^{\frac{1}{2}}[1+(1+\frac{1}{u})^{-1/2}]^{-2(s-1)} \\ &= \frac{[\sigma^{1/2} + (\sigma-1)^{1/2}]^{-2(s-1)}}{8\pi\sigma^{1/2}(\sigma-1)^{1/2}} \end{aligned} \tag{2.5}$$

for $\sigma = u + 1$. Then $G_0(p_1, p_2; s) \overset{def.}{=} \phi_s(u(p_1, p_2))$ is given by

$$G_0(p_1, p_2; s) = \frac{1}{8\pi} \frac{[\sigma(p_1, p_2)^{\frac{1}{2}} + (\sigma(p_1, p_2) - 1)^{\frac{1}{2}}]^{-2s+2}}{[\sigma(p_1, p_2)(\sigma(p_1, p_2) - 1)]^{1/2}]} \tag{2.6}$$

for $p_1 \neq p_2$. In terms of the function ϕ_s one describes the inverse of the operator - $(L + s(2 - s)1)$, as follows. Let $R(s)$ be the integral operator on $\mathbb{H}^3$ given by

$$\begin{aligned} (R(s)f)(p) &= \iiint_{\mathbb{H}^3} \phi_s(u(p, p'))f(p')dv(p') \\ &= \iiint_{\mathbb{H}^3} G_0(p, p'; s)f(p')dv(p') \end{aligned} \tag{2.7}$$

for dv in (2.2). Then for $Res > 1$, $R(s)$ is the <u>resolvent</u> of - $(L + s(2 - s)1)$- i.e. $R(s)(-L - s(2 - s)1)f = (-L - s(2 - s)1)R(s)f = f$, at least for $f \in C_c^\infty(\mathbb{H}^3)$; see, for example, [9].

We note that one can, of course, construct the resolvent kernel $G_0^{(n)}(p_1, p_2; s)$ on hyperbolic space $\mathbb{H}^{n+1}$ of arbitrary dimension (again in terms of hyperbolic distance):

$$G_0^{(n)}(p_1, p_2; s) = \pi^{-n/2} 2^{-2s-1} \frac{\Gamma(s)\sigma(p_1,p_2)^{-s}}{\Gamma(s-\frac{n}{2}+1)} \cdot \tag{2.8}$$
$$ {}_2F_1\left(s, s - \left(\tfrac{n-1}{2}\right); 2s - n + 1; \sigma(d_1, d_2)^{-1}\right).$$

For $n = 2$, $G_0^{(2)} = G_0$ by standard properties of the hypergeometric function ${}_2F_1$ [7]. For a later purpose, we note also that one can write

$$G_0(p_1, p_2; s) = \frac{1}{4\pi} \frac{e^{-d(p_1,p_2)(s-1)}}{\sinh(d(p_1, p_2))} \tag{2.9}$$

for $p_1 \neq p_2$, by an elementary manipulation of equation (2.6).

3. The heat kernel is connected to the resolvent kernel; Patterson's formula

Fix $(a, b, \alpha) \in \mathbb{R}^3$ and set

$$\gamma = \gamma_{(a,b)} \stackrel{def.}{=} \begin{bmatrix} e^{a+ib} & 0 \\ 0 & e^{-(a+ib)} \end{bmatrix}, \gamma_\alpha \stackrel{def.}{=} \begin{bmatrix} e^{i\pi\alpha} & 0 \\ 0 & e^{-i\pi\alpha} \end{bmatrix} \tag{3.1}$$

and

$$\Gamma = \Gamma_{(a,b)} \stackrel{def.}{=} \{\gamma_{(a,b)}^n \mid n \in \mathbb{Z}\}, \Gamma_{(a,b)}^{(\alpha)} \stackrel{def.}{=} \{\gamma_{(a,b)}^n \gamma_\alpha^m \mid n, m \in \mathbb{Z}\}, \tag{3.2}$$

which are the subgroups of $G = SL(2, \mathbb{C})$ generated by $\gamma_{(a,b)}$ and by $\{\gamma_{(a,b)}, \gamma_\alpha\}$, respectively. $\Gamma_{(a,b)}^{(\alpha)} = \Gamma$ for $\alpha = 1$. Later we will mainly be interested in the restrictions $a > 0$, $0 < \alpha \leq 1$. If we denote elements in $\mathbb{R}^3$ or $\mathbb{H}^3$ also by column vectors, then the standard linear fractional action of G on $\mathbb{H}^3$ restricts to the action of $\Gamma_{(a,b)}^{(\alpha)}$ on $\mathbb{H}^3$ given by

$$\left(\gamma_{(a,b)}^n \gamma_\alpha^m\right) \cdot \begin{bmatrix} x \\ y \\ z \end{bmatrix} = \begin{bmatrix} e^{2an}\{x \cos 2(bn + \pi\alpha m) - y \sin 2(bn + \pi\alpha m)\} \\ e^{2an}\{x \sin 2(bn + \pi\alpha m) + y \cos 2(bn + \pi\alpha m)\} \\ e^{2an} z \end{bmatrix} \tag{3.3}$$

for $z > 0$. In particular one obtains the action of Γ on $\mathbb{H}^3$ by taking $m = 0$ in (3.3), and one checks that a fundamental domain for this Γ-action is given by

$$F = F_{(a,b)} = \{(x, y, z) \in \mathbb{H}^3 \mid 1 < \sqrt{x^2 + y^2 + z^2} < e^{2a}\} \tag{3.4}$$

for $a > 0$. It follows that Γ is Kleinian - i.e. (by spherical coördinates)

$$vol(F) = \iiint_F dv = \infty \tag{3.5}$$

for dv in (2.2); see equation (4.7) below.

The heat kernel K_t^Γ $(t > 0)$ of $X_\Gamma \stackrel{def.}{=} \Gamma \backslash \mathbb{H}^3$ is obtained by averaging over Γ the heat kernel

$$K_t(p_1, p_2) = \frac{e^{-t-d(p_1,p_2)^2/4t} d(p_1, p_2)}{(4\pi t)^{3/2} \sinh(d(p_1, p_2))} \tag{3.6}$$

of $\mathbb{H}^3$:

$$\begin{aligned} K_t^{\Gamma}(\widetilde{p_1},\widetilde{p_2}) &= \sum_{n\in\mathbb{Z}} K_t(p_1,\gamma^n\cdot p_2) \\ &= \frac{1}{(4\pi t)^{3/2}}\sum_{n\in\mathbb{Z}}\frac{e^{-t-d(p_1,\gamma^n\cdot p_2)^2/4t}d(p_1,\gamma^n\cdot p_2)}{\sinh(d(p_1,\gamma^n\cdot p_2))}, \end{aligned} \tag{3.7}$$

by (3.2), where $\widetilde{p}$ is the Γ-orbit of $p\in\mathbb{H}^3$. The term for $n=0$ in (3.7) for $p_1=p_2$ is $(4\pi t)^{-3/2}e^{-t}$, which is <u>independent</u> of p_1, which means that the integral of $p\to K_t^{\Gamma}(\widetilde{p},\widetilde{p})$ over F <u>diverges</u> because of (3.5). One obtains a convergent integral over F by considering instead the <u>truncated</u> heat kernel

$$K_t^{*\Gamma}(\widetilde{p_1},\widetilde{p_2}) = \frac{1}{(4\pi t)^{3/2}}\sum_{n\in\mathbb{Z}-\{0\}}\frac{e^{-t-d(p_1,\gamma^n\cdot p_2)^2/4t}d(p_1,\gamma^n\cdot p_2)}{\sinh(d(p_1,\gamma^n\cdot p_2))} \tag{3.8}$$

of X_Γ, in terms of which we define the <u>theta function</u> (or heat trace)

$$trK_t^{*\Gamma} \overset{def.}{=} \iiint_F K_t^{*\Gamma}(\widetilde{p},\widetilde{p})dv(p) \tag{3.9}$$

on $\mathbb{R}^+$. One can commute the integration in (3.9) with the summation in (3.8) and obtain a concrete formula for $trK_t^{*\Gamma}$. The result is the following <u>trace formula</u> proved in [3], for example.

THEOREM 1.

$$\begin{aligned} trK_t^{*\Gamma} &= \frac{a}{2\sqrt{4\pi t}}\sum_{n\in\mathbb{Z}-\{0\}}\frac{e^{-\left(t+\frac{n^2a^2}{t}\right)}}{[\sinh^2(na)+\sin^2(nb)]} \\ &= \frac{a}{\sqrt{4\pi t}}\sum_{n=1}^{\infty}\frac{e^{-\left(t+\frac{n^2a^2}{t}\right)}}{[\sinh^2(na)+\sin^2(nb)]}. \end{aligned} \tag{3.10}$$

Compare also [2,5,10,13].

To connect the heat kernel $K_t^{*\Gamma}$ with the resolvent kernel G_0 (or <u>Green's function</u>) of section 2, we consider the following transform

$$I(s) \overset{def.}{=} \int_0^\infty e^{-s(s-2)t}trK_t^{*\Gamma}dt \tag{3.11}$$

of $K_t^{*\Gamma}$. Commuting the integration in (3.11) with the integration in (3.9) (which can be justified for $Res>1$) one obtains by (3.8)

$$I(s) = \iiint_F \sum_{n\neq 0}\frac{a_n}{\sinh a_n}\left[\int_0^\infty \frac{e^{-s(s-2)t}e^{-t-a_n^2/4t}dt}{(4\pi t)^{3/2}}\right]dv(p) \tag{3.12}$$

for $a_n = a_n(p) \overset{def.}{=} d(p,\gamma^n\cdot p)$, where the integral $\int_0^\infty$ in (3.12) has the value $2\pi^{1/2}e^{-a_n(s-1)}/(4\pi)^{3/2}a_n$ for $Res>1$ by page 1145 of [7]. It follows that

$$\begin{aligned} I(s) &= \frac{1}{4\pi}\iiint_F \sum_{n\in\mathbb{Z}-\{0\}}\frac{e^{-d(p,\gamma^n\cdot p)(s-1)}}{\sinh(d(p,\gamma^n\cdot p))}dv(p) \\ &= \iiint_F \sum_{n\in\mathbb{Z}-\{0\}} G_0(p,\gamma^n\cdot p;s)dv(p), \end{aligned} \tag{3.13}$$

by (2.9). Here we note that on the fundamental domain F, $\gamma^n \cdot p_1 \neq p_2$ for $n \neq 0$; in particular $n \neq 0 \Rightarrow a_n \neq 0 \Rightarrow p \neq \gamma^n \cdot p$ for $p \in F$.

On the other hand, by Theorem 1

$$I(s) \overset{def.}{=} \int_0^\infty e^{-s(s-2)t} \frac{a}{\sqrt{4\pi t}} \sum_{n=1}^\infty \frac{e^{-\left(t+\frac{n^2a^2}{t}\right)} dt}{[\sinh^2(na)+\sin^2(nb)]}. \tag{3.14}$$

By commuting the integration and summation in (3.14) and applying, again, an integral formula on page 1145 of [7], for $Res > 1$, one concludes that

$$\begin{aligned} I(s) &= \frac{a}{2(s-1)} \sum_{n=1}^\infty \frac{e^{-2an(s-1)}}{[\sinh^2(an)+\sin^2(bn)]} \\ &\equiv \frac{1}{s-1} Z'_\Gamma(s)/Z_\Gamma(s) \end{aligned} \tag{3.15}$$

where

$$Z_\Gamma(s) \overset{def.}{=} \Pi^\infty_{0\leq k_1,k_2\in\mathbb{Z}} \left[1-(e^{i2b})^{k_1}(e^{-i2b})^{k_2} e^{-(k_1+k_2+s)2a}\right] \tag{3.16}$$

is Patterson's Selberg zeta function attached to Γ [11]. In other words, by (3.13) and (3.15) we obtain, by an alternate approach, Patterson's formula (for $Res > 1$)

$$\iiint_F G_0^\Gamma(p,p;s)dv(p) = \frac{1}{s-1} Z'_\Gamma(s)/Z_\Gamma(s) \tag{3.17}$$

where

$$G_0^\Gamma(p_1,p_2;s) \overset{def.}{=} \sum_{n\in\mathbb{Z}-\{0\}} G_0(p_1,\gamma^n \cdot p_2;s) \tag{3.18}$$

serves as the resolvent kernel of X_Γ. Here our normalization of the zeta function $Z_\Gamma(s)$ in (3.16) differs (by a factor of 2) from that in [11]. Going back to definition (3.11) we have also shown that for $Res > 1$

$$\int_0^\infty e^{-s(s-2)t} tr K_t^{*\Gamma} dt = \frac{1}{s-1} Z'_\Gamma(s)/Z_\Gamma(s), \tag{3.19}$$

which is a formula analogous to Gangolli's formula (2.38) in [6], where ρ_0 there has the value 1 for our symmetric space $SL(2,\mathbb{C})/SU(2) = \mathbb{H}^3$. Of course the Γ in [6] is co-compact, and therefore is certainly not Kleinian: (3.5) cannot hold in the co-compact (or co-finite volume) case. We see also a connection between the heat and resolvent kernels $K_t^{*\Gamma}$, G_0^Γ:

$$\int_0^\infty e^{-s(s-2)t} tr K_t^{*\Gamma} dt = \iiint_F G_0^\Gamma(p,p;s)dv(p) \tag{3.20}$$

for $Res > 1$, the common value being given by (3.19).

4. An extension of Patterson's formula

Assume henceforth that $a > 0$ and $0 < \alpha \leq 1$. The subgroup $\Gamma^{(\alpha)}_{(a,b)}$ in (3.2) is the $\Gamma^{(\alpha)}$ mentioned in the introduction, and the deformation $\{Z_{\Gamma^{(\alpha)}}(s)\}_{0<\alpha\leq 1}$ mentioned there is given in [17, 18, 19] by

$$Z_{\Gamma^{(\alpha)}_{(a,b)}}(s) = Z(s;\alpha,a,b) \overset{def.}{=}$$

$$\Pi^{\infty}_{0\leq k_1,k_2\in\mathbb{Z}}\left[1-(e^{\frac{i2b}{\alpha}})^{k_1}(e^{-\frac{i2b}{\alpha}})^{k_2}e^{-(k_1+k_2+\alpha s)\frac{2a}{\alpha}}\right]\cdot$$

$$\Pi^{\infty}_{0\leq k_1,k_2,k_3\in\mathbb{Z}}\frac{\left[1-e^{-2(k_3+1)2a}(e^{\frac{i2b}{\alpha}})^{k_1}(e^{-\frac{i2b}{\alpha}})^{k_2}e^{-(k_1+k_2+\alpha s)\frac{2a}{\alpha}}\right]}{\left[1-e^{-2(k_3+\frac{1}{\alpha})2a}(e^{\frac{i2b}{\alpha}})^{k_1}(e^{-\frac{i2b}{\alpha}})^{k_2}e^{-(k_1+k_2+\alpha s)\frac{2a}{\alpha}}\right]}. \tag{4.1}$$

Thus clearly $Z_{\Gamma^{(1)}_{(a,b)}}(s) = Z_{\Gamma_{(a,b)}}(s)$, by (3.16). A second (simpler) expression

$$\begin{aligned} Z(s;\alpha,a,b) &= \Pi^{\infty}_{0\leq k_1,k_2\in\mathbb{Z}}[1-e^{(ib-a)\frac{2k_1}{\alpha}}e^{-4ak_2}e^{-2as}]\cdot \\ &\quad \Pi^{\infty}_{0\leq k_1,k_2\in\mathbb{Z}}[1-e^{-(ib+a)\frac{2(k_1+1)}{\alpha}}e^{-4ak_2}e^{-2as}] \end{aligned} \tag{4.2}$$

shows that $Z(s;\alpha,a,b)$ is an entire function of s. One has that

$$\frac{Z'(s;\alpha,a,b)}{Z(s;\alpha,a,b)} = a\sum_{n=1}^{\infty}\frac{\sinh(\frac{2an}{\alpha})e^{-(s-1)2an}}{\sinh(2an)[\cosh(\frac{2an}{\alpha})-\cos(\frac{2bn}{\alpha})]} \tag{4.3}$$

for $Re s > 0$.

We also assume now that $\alpha = \frac{1}{l}$ for an integer $l > 1$. From (3.3) the action of the subgroup $\{\gamma^m_\alpha \mid m \in \mathbb{Z}\}$ of $\Gamma^{(\alpha)}_{(a,b)}$ on $\mathbb{H}^3$ amounts to the action of $\mathbb{Z}$ on $\mathbb{R}^2$ given by a notation through the angle $2\pi\alpha m$

$$m\cdot\begin{bmatrix} x \\ y\end{bmatrix} \overset{def.}{=} \begin{bmatrix}\cos(2\pi\alpha m) & -\sin(2\pi\alpha m)\\ \sin(2\pi\alpha m) & \cos(2\pi\alpha m)\end{bmatrix}\begin{bmatrix} x \\ y\end{bmatrix}, \tag{4.4}$$

which for $\alpha = \frac{1}{l}$ has a fundamental domain that is a cone in $\mathbb{R}^2$ with the origin as its vertex, and with an opening angle $2\pi/l = 2\pi\alpha$. The defect angle spoken of in the introduction is the angle $2\pi - 2\pi\alpha = 2\pi(1-\alpha)$. The quotients $X_{\Gamma^{(\alpha)}_{(a,b)}} \overset{def.}{=} \Gamma^{(\alpha)}_{(a,b)}\backslash\mathbb{H}^3$ therefore have conical singularities. They are not smooth manifolds: $(0,0,z)$, $z > 0$, for example, are in fact $\Gamma^{(\alpha)}_{(a,b)}$-fixed points for $n = 0$, by (3.3). Note also that the action of $\mathbb{Z}$ on $\mathbb{R}^2$ in (4.4) projects to an action of the finite group $\mathbb{Z}_l \overset{def.}{=} l\mathbb{Z}\backslash\mathbb{Z}$ on $\mathbb{R}^2$, since $m - m_1 \in l\mathbb{Z} \Rightarrow \cos(2\pi\alpha m) = \cos(2\pi\alpha m_1)$, $\sin(2\pi\alpha m) = \sin(2\pi\alpha m_1)$ for $\alpha = 1/l$. Thus we can average $K^{*\Gamma}_t$ in (3.8) over $\mathbb{Z}_l$ and G^{Γ}_0 in (3.18) and obtain

$$K^{*(\alpha)}_t(\widetilde{p_1},\widetilde{p_2}) \overset{def.}{=} \sum_{m=0}^{l-1} K^{*\Gamma}_t(\widetilde{p_1},\widetilde{\gamma^{(\alpha)}_m\cdot p_2}) =$$

$$\frac{1}{(4\pi t)^{3/2}}\sum_{m=0}^{l-1}\sum_{n\in\mathbb{Z}-\{0\}}\frac{e^{-t-d(p_1,(\gamma^n\gamma^{(\alpha)}_m)\cdot p_2)^2/4t}d(p_1,(\gamma^n\gamma^{(\alpha)}_m)\cdot p_2)}{\sinh(d(p_1,(\gamma^n,\gamma^{(\alpha)}_m)\cdot p_2))}, \tag{4.5}$$

$$
\begin{aligned}
G_0^{\Gamma^{(\alpha)}}(p_1, p_2; s) &\overset{def.}{=} \sum_{m=0}^{l-1} G_0^{\Gamma}(p_1, \gamma_m^{(\alpha)} \cdot p_2; s) \\
&= \sum_{m=0}^{l-1} \sum_{n\in\mathbb{Z}-\{0\}} G_0(p_1, (\gamma^n \gamma_m^{(\alpha)})) \cdot p_2; s).
\end{aligned} \tag{4.6}
$$

Using spherical coördinates $\rho \geq 0$, $0 \leq \theta < 2\pi$, $0 \leq \phi < \frac{\pi}{2}$, $x = \rho \sin\phi \cos\theta$, $y = \rho \sin\phi \sin\theta$, $z = \rho\cos\phi$, one notes that for a function f on F,

$$
\begin{aligned}
&\iiint_F f(p) dv(p) = \\
&\int_0^{\pi/2} \int_0^{2\pi} \int_1^{e^{2a}} f(\rho\sin\phi\cos\theta, \rho\sin\phi\sin\theta, \rho\cos\theta) \frac{\sin\phi}{\rho\cos^3\phi} d\rho d\theta d\phi,
\end{aligned} \tag{4.7}
$$

given (2.2), (3.4). One can generalize the arguments in [3] for the proof of Theorem 1 and show that

$$
\begin{aligned}
&\int_0^{\pi/2} \int_{\theta_1}^{\theta_2} \int_1^{e^{2a}} K_t^{*(\alpha)}(\widetilde{p}, \widetilde{p}) dv(p) = \\
&\frac{(\theta_2 - \theta_1)}{\sqrt{4\pi t}} \left(\frac{a}{4\pi}\right) \sum_{\substack{n\in\mathbb{Z}-\{0\} \\ 0\leq m\leq l-1}} \frac{e^{-t-a^2n^2/t}}{[\sinh^2(an) + \sin^2(bn + \pi\alpha m)]}.
\end{aligned} \tag{4.8}
$$

Of course for $l = 1$ (i.e. $\alpha = 1$), $\theta_1 = 0$, and $\theta_2 = 2\pi$, formula (4.8) reduces to Theorem 1.

Following the arguments of section 3, one computes (for $Res > 1$) that

$$
\begin{aligned}
&\int_0^\infty e^{-s(s-2)t}[\text{the r.h.s. of (4.8)}] dt = \\
&\frac{a(\theta_2 - \theta_1)}{4\pi(s-1)} \sum_{\substack{n\neq 0 \\ 0\leq m\leq l-1}} \frac{e^{-2a|n|(s-1)}}{2[\sinh^2(an) + \sin^2(bn + \pi\alpha m)]} = \\
&\frac{a(\theta_2 - \theta_1)}{4\pi(s-1)} \sum_{\substack{n\neq 0 \\ 0\leq m\leq l-1}} \frac{e^{-2a|n|(s-1)}}{\cosh(2an) - \cos 2(bn + \pi\alpha m)},
\end{aligned} \tag{4.9}
$$

which puts one in position to use the identity

$$
\sum_{m=0}^{l-1} \frac{1}{\cosh u - \cos(v - \frac{2\pi m}{l})} = \frac{l \sin(lu)}{(\sinh u)(\cosh lu - \cos lv)}. \tag{4.10}
$$

Thus for the choices $u = 2an$, $v = 2bn$, one sees that

$$\int_0^\infty e^{-s(s-2)t}[\text{the r.h.s. of (4.8)}]dt =$$

$$\frac{a(\theta_2-\theta_1)}{4\pi(s-1)} \sum_{n\in\mathbb{Z}-\{0\}} \frac{l\sinh(2anl)e^{-2a|n|(s-1)}}{\sinh(2an)[\cosh(2anl)-\cos(2bnl)]}$$

$$= \frac{a(\theta_2-\theta_1)}{4\pi(s-1)} 2\sum_{n=1}^\infty \frac{l\sinh(2anl)e^{-2an(s-1)}}{\sinh(2an)[\cosh(2anl)-\cos(2bnl)]}$$

$$= \frac{(\theta_2-\theta_1)2}{4\pi\alpha(s-1)} \frac{Z'(s;\alpha,a,b)}{Z(s;\alpha,a,b)} \tag{4.11}$$

by (4.3).

On the other hand, by definition (4.5)

$$\int_0^\infty e^{-s(s-2)t}[\text{the l.h.s. of (4.8) }]dt = \tag{4.12}$$

$$\int_0^{\pi/2}\int_{\theta_1}^{\theta_2}\int_1^{e^{2a}} \sum_{\substack{n\neq 0\\ 0\le m\le l-1}} \frac{a_{nm}}{\sinh a_{nm}} \left[\int_0^\infty \frac{e^{-s(s-2)t}e^{-t-a_{nm}^2/4t}dt}{(4\pi t)^{3/2}}\right] dv(p)$$

for $a_{nm} = a_{nm}(p) \overset{def.}{=} d(p,(\gamma^n\gamma_m^{(\alpha)})\cdot p)$, $Res > 1$, (similarly to equation (3.12)), where as above $[\int_0^\infty]=\frac{2\pi^{1/2}e^{-a_{nm}(s-1)}}{(4\pi)^{3/2}a_{nm}}$. This gives that (by (2.9), (3.18), (4.6))

$$\int_0^\infty e^{-s(s-2)t}[\text{the l.h.s. of (4.8)}]dt =$$

$$\int_0^{\pi/2}\int_{\theta_1}^{\theta_2}\int_1^{e^{2a}} \sum_{\substack{0\le m\le l-1\\ n\neq 0}} \frac{2\pi^{1/2}e^{-a_{nm}(p)(s-1)}}{(4\pi)^{3/2}\sinh a_{nm}(p)} dv(p) = \tag{4.13}$$

$$\int_0^{\pi/2}\int_{\theta_1}^{\theta_2}\int_1^{e^{2a}} \sum_{0\le m\le l-1}\sum_{n\neq 0} G_0(p,(\gamma^n\gamma_m^{(\alpha)})\cdot p;s)dv(p) =$$

$$\int_0^{\pi/2}\int_{\theta_1}^{\theta_2}\int_1^{e^{2a}} G_0^{\Gamma^{(\alpha)}}(p,p;s)dv(p).$$

By equations (4.11), (4.13) we get

$$\int_0^{\pi/2}\int_{\theta_1}^{\theta_2}\int_1^{e^{2a}} G_0^{\Gamma^{(\alpha)}}(p,p;s)dv(p) =$$

$$\frac{(\theta_2-\theta_1)2}{4\pi\alpha(s-1)}\frac{Z'(s;\alpha,a,b)}{Z(s;\alpha,a,b)} \tag{4.14}$$

for $Res > 1$.

Denote by $F^{(\alpha)}$ the upper hemispherical region in $\mathbb{R}^3$ between the spheres $\rho = 1$, $\rho = e^{2a}$, but with θ at least the value of the defect angle:

$$F^{(\alpha)}: \qquad 1 < \rho < e^{2a}$$

$$2\pi(1-\alpha) \le \theta \le 2\pi \tag{4.15}$$

For the choices $\theta_1 = 2\pi(1-\alpha)$, $\theta_2 = 2\pi$ in (4.14) (so that $\theta_2 - \theta_1 = 2\pi\alpha$) we see that the following extension of Patterson's formula (formula (3.17)) holds.

THEOREM 2. *(Resolvent trace formula in the presence of a conical singularity)*

$$\iiint_{F^{(\alpha)}} G_0^{\Gamma^{(\alpha)}}(p,p;s)dv(p) =$$

$$\frac{1}{s-1}Z'(s;\alpha,a,b)/Z(s;\alpha,a,b) \tag{4.16}$$

for $Res > 1$, $a > 0$, $F^{(\alpha)}$ *in (4.15),* $G_0^{\Gamma^{(\alpha)}}$ *in (4.6),* $Z(s;\alpha,a,b)$ *in (4.1) (or in (4.2)),* $1/\alpha \in \mathbb{Z}$*; also see (4.3).*

For $\alpha = 1$, $F^{(\alpha)} = F$ in (3.4) and formula (4.16) reduces to formula (3.17).

5. BTZ

Since we have not specified any choice of a, b in definition (3.1), but have worked in full generality, the discussion so far has been purely mathematical. We indicate now, briefly, which choice of a, b renders $X_\Gamma \stackrel{def.}{=} \Gamma_{(a,b)}\backslash\mathbb{H}^3$ a black hole solution of Einstein's equation, and which therefore renders results of the previous sections of some physical interest. Further details appear in [1, 2, 4, 10, 14, 15, 16, 17, 18].

An interesting and important three-dimensional black hole solution, with a negative cosmological constant Λ, of the Einstein gravitational, vacuum field equations

$$R_{ij} - \frac{g_{ij}}{2}R - \Lambda g_{ij} = 0 \tag{5.1}$$

was found in 1992 by M. Bañados, C. Teitelboim, and J. Zanelli [1], which we have referred to as the BTZ black hole. Here R_{ij}, R are the Ricci tensor and scalar curvature, respectively, of the metric g_{ij}, which lives on a portion of anti-deSitter space with variables (r, ϕ, τ) where r is radial, ϕ is a periodic Schwarzschild variable, and τ is a "time" variable. The scalar curvature is constant and is given in terms of the cosmological constant: $R = 6/\sigma^2$ for $\sigma \stackrel{def.}{=} 1/\sqrt{-\Lambda}$. It is a very nice fact that, due to Schwarzschild periodicity of ϕ, one can set up a change of variables $(r, \phi, \tau) \to (x, y, z)$ such that the BTZ metric g_{ij} assumes the standard hyperbolic form (2.1), up to multiplication by σ^2 [4]:

$$ds^2_{BTZ} = \frac{\sigma^2}{z^2}(dx^2 + dy^2 + dz^2). \tag{5.2}$$

Moreover the topology of the black hole, whose Euclidean version we consider (there is a also a Lorentzian version), is given by the quotient $X_\Gamma = \Gamma_{(a,b)}\backslash\mathbb{H}^3$ for a suitable choice of a, b in (3.2) that we now describe. X_Γ has outer and inner horizons are given, respectively, by $r_+ > 0$, $r_- \in i\mathbb{R}$ $(i^2 = -1)$ where

$$\begin{aligned} r_+^2 &= \frac{M\sigma^2}{2}[1 + (1 + \frac{J^2}{M^2\sigma^2})^{1/2}] \\ r_- &= -\frac{\sigma J i}{2r_+} \end{aligned} \tag{5.3}$$

where $M > 0$, $J \geq 0$ are its mass and angular momentum, respectively. The main point is that the suitable choice of a, b is

$$a \stackrel{def.}{=} \pi r_+/\sigma > 0, b \stackrel{def.}{=} \pi|r_-|/\sigma = \frac{\pi J}{2r_+} \geq 0. \tag{5.4}$$

In section 4, the main section, we did assume that a was positive, which was needed for the assertion (3.4), for example, or for the definition of the domain $F^{(\alpha)}$ in (4.15). One can allow for $b = 0$ - i.e. for $J = 0$ - i.e. for the special case of a non-spinning black hole.

6. Appendices

Following the referee's suggestion, who we thank for a careful review of the paper and for helpful corrections, we offer for the reader's benefit some further remarks (expository in nature) regarding some of the definitions employed in previous sections, and some proofs.

A1. The volume element in (2.2)

The Poincaré metric in (2.1) is expressed alternately as

$$g = g(x,y,z) = \begin{bmatrix} z^{-2} & 0 & 0 \\ 0 & z^{-2} & 0 \\ 0 & 0 & z^{-2} \end{bmatrix} \tag{6.1}$$

for $(x,y,z) \in \mathbb{H}^3$. By definition, the Riemannian volume dv corresponding to the Riemannian metric g is given by

$$dv \overset{def.}{=} \sqrt{|det g|}dxdydz \doteq z^{-3}dxdydz. \tag{6.2}$$

which gives (2.2).

By taking the function f in equation (4.7) equal to 1, one sees that indeed $vol(F) = \int_0^{\pi/2}\int_0^{2\pi}\int_1^{e^{2a}} (\sin\phi)\rho^{-1}\cos^{-3}\phi d\rho d\theta d\phi = (2a)2\pi\left[\frac{\cos^{-2}\phi}{2}\right]_{\phi=0}^{\phi=\pi/2}$ is infinite, as remarked in (3.5).

A2. $\mathbb{H}^3$ described as the homogeneous space $SL(2,\mathbb{C})/SU(2)$

It is well-known that hyperbolic 2-space $\mathbb{H}^2 \overset{def.}{=} \{(x,y) \in \mathbb{R}^2 \mid y > 0\}$ (i.e. the upper $\frac{1}{2}$-plane) can be described as the homogeneous space G_0/K_0 for $G_0 = SL(2,\mathbb{R})$, $K_0 = SO(2)$. That is, the standard linear fractional action $(g_0, z) \to g_0 \cdot z \overset{def.}{=} (az+b)(cz+d)^{-1}$ of G_0 on $\mathbb{H}^2$, for $g_0 = \begin{bmatrix} a & b \\ c & d \end{bmatrix} \in G_0$, $z \in \mathbb{H}^2$, is transitive (i.e. for any two points $p_1, p_2 \in \mathbb{H}^2$ one can find some $g_0 \in G_0$ such that $p_2 = g_0 \cdot p_1$) and it has K_0 as the isotropy subgroup of G_0 at the point $p_0 \overset{def.}{=} (0,1) = i \in \mathbb{H}^2$ (i.e. $K_0 \equiv \{g_0 \in G_0 \mid g_0 \cdot i = i\}$). In a similar way, one has a standard linear fractional action of $G = SL(2,\mathbb{C})$ on $\mathbb{H}^3$; see [19], for example, for details. For $(g,p) = \left(\begin{bmatrix} a & b \\ c & d \end{bmatrix}, \begin{bmatrix} x \\ y \\ z \end{bmatrix}\right) \in G \times \mathbb{H}^3$, the equations $g \cdot p \overset{def.}{=} (u,v,w) \in \mathbb{H}^3$ for

$$\begin{aligned} u + iv &\overset{def.}{=} [(at+b)\overline{(ct+d)} + a\bar{c}z^2][|ct+d|^2 + |c|^2z^2]^{-1}, \\ w &\overset{def.}{=} z\left[|ct+d|^2 + |c|^2z^2\right]^{-1}, \qquad t \overset{def.}{=} x + iy \end{aligned} \tag{6.3}$$

explicitly describe the action, where the bar "-" denotes complex conjugation. In particular, equation (3.3) follows from (6.3). We check that, similar to the case of G_0 acting on $\mathbb{H}^2$, the action of G on $\mathbb{H}^3$ is transitive, and we compute the isotropy subgroup

$$G_{p_0} \overset{def.}{=} \{g \in G \mid g \cdot p_0 = p_0\} \tag{6.4}$$

of G at the point $p_0 \overset{def.}{=} \begin{bmatrix} 0 \\ 0 \\ 1 \end{bmatrix} \in \mathbb{H}^3$, showing that it is the subgroup $SU(2)$.

In fact let $p_j = \begin{bmatrix} x_j \\ y_j \\ z_j \end{bmatrix} \in \mathbb{H}^3$ be given, $j = 1, 2$. Define

$$g_j = \begin{bmatrix} (x_j + iy_j)z_j^{-1/2} & -z_j^{1/2} \\ z_j^{-1/2} & 0 \end{bmatrix} \in G. \tag{6.5}$$

Then $g_j \cdot p_0 = (u_j, v_j, w_j) \in \mathbb{H}^3$ where, by (6.3), $u_j + iv_j = x_j + iy_j$, $w_j = z_j$. That is, $g_j \cdot p_0 = p_j \Rightarrow p_2 = (g_2 g_1^{-1}) \cdot p_1$, which proves transitivity of the G-action. Also by (6.3), $g \cdot p_0 = (u, v, w)$ for $w = [|c|^2 + |d|^2]^{-1}$, $u + iv = [b\bar{d} + a\bar{c}][|c|^2 + |d|^2]^{-1}$. Thus $g \cdot p_0 = p_0$ (i.e. $g \in G_{p_0}$ in (6.4)) $\Leftrightarrow u = v = 0, w = 1 \Leftrightarrow |c|^2 + |d|^2 = 1, b\bar{d} + a\bar{c} = 0$, in which case one notes that for $g^* \overset{def.}{=} \bar{g}^t =$ the adjoint of g one has that

$$gg^* = \begin{bmatrix} |a|^2 + |b|^2 & 0 \\ 0 & 1 \end{bmatrix}. \tag{6.6}$$

Since $detgg^* = 1$ (as $g \in G$), one concludes that $|a|^2 + |b|^2 = 1$. That is, $gg^* = 1 \Rightarrow g \in SU(2)$, which proves that $G_{p_0} = SU(2)$. It follows, by general principles, that the map $gSU(2) \to g \cdot p_0$ is a well-defined homeomorphism (or even a diffeomorphism) of $SL(2, \mathbb{C})/SU(2)$ onto $\mathbb{H}^3$.

The identification $SL(2, \mathbb{C})/SU(2) = \mathbb{H}^3$ also follows, of course, from the fact that $SL(2, \mathbb{C})$ has an Iwasawa decomposition with respect to the maximal compact subgroup $SU(2)$. Following the agreement made in section 3, we have continued to denote elements in $\mathbb{R}^3$ or in $\mathbb{H}^3$ by both row and column vectors.

A3. F is a fundamental domain

In section 3. it was remarked that one could check that the subset $F = F_{(a,b)}$ defined in (3.4) for $a > 0$ (which is independent of $b \in \mathbb{R}$) is a fundamental domain for the action of $\Gamma = \Gamma_{(a,b)}$ (see (3.2)) on $\mathbb{H}^3$. By definition this means $F \subset \mathbb{H}^3$ is an open subset such that (i) given $p_1, p_2 \in F$ with $p_1 \neq p_2$, the equation $g \cdot p_1 = p_2$ has no solution $g \in \Gamma$ (thus no two distinct points of Γ lie in the same Γ-orbit) (ii) given $p \in \mathbb{H}^3$, $\exists g \in \Gamma$ such that $g \cdot p \in \overline{F}$ (= the closure of F). Here of course we are restricting to Γ the action of $G = SL(2, \mathbb{C})$ (discussed in A2.) on $\mathbb{H}^3$. We provide a proof that conditions (i),(ii) indeed hold. Since $\mathbb{H}^3$ has the relative topology of $\mathbb{R}^3$, $F \subset \mathbb{H}^3$ is certainly an open subset. In fact $F \overset{def.}{=}$ the inverse image $\rho^{-1}((1, N))$ of the open interval $(1, N) \subset \mathbb{R}$ for $N \overset{def.}{=} e^{2a} > 1$ (as $a > 0$) and for $\rho : \mathbb{H}^3 \to \mathbb{R}$ given by $\rho(x, y, z) \overset{def.}{=} +\sqrt{x^2 + y^2 + z^2}$, which is positive as $z > 0$.

Suppose $p_j = (x_j, y_j, z_j) \in \mathbb{H}^3$, $j = 1, 2$, lie in the same Γ-orbit: $p_2 = g \cdot p_1$ for some $g = \gamma^n_{(a,b)} \in \Gamma$, $n \in \mathbb{Z}$. By (3.3) (take $m = 0$ there), $z_2 = e^{2an} z_1, y_2 = e^{2an}(x_1 \sin 2bn + y_1 \cos 2bn), x_2 = e^{2an}(x_1 \cos 2bn - y_1 \sin 2bn)$, so that for $\rho_j \overset{def.}{=} \rho(x_j, y_j, z_j) > 0$ a quick computation gives $\rho_2^2 = (e^{2an})^2 \rho_1^2$, or $\rho_2 \overset{a.}{=} e^{2an}\rho_1$. In particular if $p_1, p_2 \in F = \rho^{-1}((1, N))$ we have that $1 < \rho_1, \rho_2 < N \Rightarrow 0 < \log \rho_1, \log \rho_2 < \log N \overset{def.}{=} 2a$, where by $a.$, $\log \rho_2 = 2an + \log \rho_1$. That is, $2a > |\log \rho_1 - \log \rho_2| = 2a|n| \Rightarrow |n| < 1 \Rightarrow n = 0 \Rightarrow g = 1 \Rightarrow p_2 = p_1$, which establishes condition (i). To prove condition (ii) let $p_1 = (x_1, y_1, z_1) \in \mathbb{H}^3$ be an arbitrary point. Given $t \in \mathbb{R}$, the largest integer $[t]$ that does not exceed t satisfies $t - 1 < [t] \leq t$. For the choice $t \overset{def.}{=} 1 - (\log \rho_1)/2a$ (again $\rho_1 \overset{def.}{=} \rho(p_1) = +\sqrt{x_1^2 + y_1^2 + z_1^2}$) we define $n = n(p_1) = [t]$ and we thus conclude that $-(\log \rho_1)/2a < n \leq 1 - (\log \rho_1)/2a$, which we multiply by $2a$ (again for $a > 0$): $-\log\rho_1 < 2an \leq 2a - \log\rho_1 \Rightarrow \frac{1}{\rho_1} < e^{2an} \leq \frac{N}{\rho_1}$ (since $N \overset{def.}{=} e^{2a}$) $\Rightarrow 1 < e^{2an}\rho_1 \leq N$. That is, define $\rho_2 \overset{def.}{=} \gamma^n \cdot p_1$ (again for γ in (3.1)) so that by equation $a.$, $\rho_2 = e^{2an}\rho_1 \in (1, N] \Rightarrow p_2 \in \rho^{-1}((1,N]) \subset \overline{F}$. That is, given any $p_1 \in \mathbb{H}^3$ we have produced an element $g = \gamma^n \in \Gamma$ such that $g \cdot p \overset{def.}{=} p_2 \in \overline{F}$, which proves condition (ii), and we have shown that the integer n, in fact, can be chosen as follows: $n =$ the largest integer that does not exceed $1 - (\log\rho(x_1, y_1, z_1))/2a$ for $p_1 = (x_1, y_1, z_1), \rho(x_1, y_1, z_1) = +\sqrt{x_1^2 + y_1^2 + z_1^2}$.

A4. More on the zeta function $\mathbf{Z}(\mathbf{s}; \alpha, \mathbf{a}, \mathbf{b})$

The zeta function $Z(s; \alpha, a, b)$, for $0 < \alpha \leq 1$, $a > 0$, $b \geq 0$, $s \in \mathbb{C}$, is defined in equation (4.1). The definition was motivated initially by physical considerations. Namely, as pointed out in [17, 18, 19] if D_α is the differential operator $\alpha\frac{\partial}{\partial\alpha} - 1$, then our goal was to construct a type of statistical mechanics function $\log Z(s; \alpha, a, b)$ such that the evaluation of $D_\alpha \log Z(s; \alpha, a, b)$ at $\alpha = 1$ would capture the Mann-Solodukhin quantum correction to BTZ black hole entropy, for a suitable value of s and for a, b specialized as in (5.4). We have claimed that $Z(s; \alpha, a, b)$ (again for arbitrary $a > 0$, $b \geq 0$) is also given by the (simpler) product expression in (4.2). It is certainly not clear (as the referee has also pointed out) that equation (4.2) indeed does hold. This we shall prove in this appendix, and in fact we shall also obtain a third (simpler) quotient expression of $Z(s; \alpha, a, b)$; see equation (6.14) below. For this purpose, it is convenient to consider the convergent products

$$G_r(z; q_1, q_2, ..., q_r) \overset{def.}{=} \Pi_{0 \leq k_1, k_2, ..., k_r \in \mathbb{Z}}[1 - q_1^{k_1} \cdots q_r^{k_r} z] \tag{6.7}$$

defined for $r = 1, 2, 3, ..., z, q_1, ..., q_r \in \mathbb{C}$, $|q_i| < 1$. Our interest is only in the cases $r = 2, 3$.

The key property of these products that we use is what we call the factor or substitution property:

$$\begin{aligned} G_3(z; q_1, q_2, q_3) &= G_2(z; q_2, q_3)G_3(q_1 z; q_1, q_2, q_3) \\ &= G_2(z; q_1, q_3)G_3(q_2 z; q_1, q_2, q_3) \\ &= G_2(z; q_1, q_2)G_3(q_3 z; q_1, q_2, q_3); \end{aligned} \tag{6.8}$$

here $q_1 z, q_2 z, q_3 z$ "substitute" for z. We prove the first factorization in (6.8) - the proof of the second and third being entirely analogous. Write

$$\begin{aligned} G_3(q_1 z; q_1, q_2, q_3) &\overset{def.}{=} \Pi_{0 \leq k, l, m \in \mathbb{Z}}[1 - q_1^k q_2^l q_3^m (q_1 z)] \\ &= \Pi_{\substack{0 \leq k \in \mathbb{Z} \\ 0 \leq l, m \in \mathbb{Z}}} [1 - q_1^{k+1} q_2^l q_3^m z] \\ &= \Pi_{\substack{1 \leq k \in \mathbb{Z} \\ 0 \leq l, m \in \mathbb{Z}}} [1 - q_1^k q_2^l q_3^m z]. \end{aligned} \tag{6.9}$$

Then for $G_3(z; q_1, q_2, q_3)$ as a product over $0 \leq k, l, m \in \mathbb{Z}$, we write out the terms for $k = 0$, $k \geq 1$:

$$\begin{aligned} G_3(z; q_1, q_2, q_3) &\overset{def.}{=} \Pi_{0 \leq k, l, m \in \mathbb{Z}}[1 - q_1^k q_2^l q_3^m z] \\ &= \Pi_{0 \leq l, m \in \mathbb{Z}}[1 - q_2^l q_3^m z] \Pi_{\substack{1 \leq k \in \mathbb{Z} \\ 0 \leq l, m \in \mathbb{Z}}} [1 - q_1^k q_2^l q_3^m z] \\ &\overset{def.}{=} G_2(z; q_2, q_3) G_3(q_1 z; q_1, q_2, q_3), \end{aligned} \tag{6.10}$$

by (6.9), which is the desired result. By considering $G_r(q_j z; q_1, q_2, ..., q_r)$ for $1 \leq j \leq r$ in place of $G_3(q_1 z; q_1, q_2, q_3)$ in (6.9), one can clearly generalize the argument just given and prove the generalization

$$G_r(q_j z; q_1, q_2, ..., q_r) = G_{r-1}(z; q_1, ..., \hat{q}_j, ..., q_r) G_r(q_j z; q_1, q_2, ..., q_r) \tag{6.11}$$

of (6.8) for $r \geq 2$, where $\hat{q}_j$ denotes the omission of q_j.

In the first equation in (6.8) replace z by $q_2 z$: $G_3(q_2 z; q_1, q_2, q_3) = G_2(q_2 z; q_2, q_3) \cdot G_3(q_1 q_2 z; q_1, q_2, q_3)$. Therefore by the second equation in (6.8)

$$G_3(z; q_1, q_2, q_3) = G_2(z; q_1, q_3) G_2(q_2 z; q_2, q_3) G_3(q_1 q_2 z; q_1, q_2, q_3). \tag{6.12}$$

The calculus of the products G_2, G_3 is now applied to our zeta function $Z(s; \alpha, a, b)$ defined in equation (4.1) for $0 < \alpha \leq 1$, $a > 0$, $b \geq 0$, $s \in \mathbb{C}$. For this, choose $q_1 \overset{def.}{=} e^{-2a/\alpha} e^{2bi/\alpha}$, $q_2 \overset{def.}{=} e^{-2a/\alpha} e^{-2bi/\alpha}$, $q_3 \overset{def.}{=} e^{-4a}$; indeed as α, b are real and $a, \alpha > 0$, $|q_j| < 1$ for $j = 1, 2, 3$. Then, by definition, one has

$$Z(s; \alpha, a, b) = \frac{G_2(e^{-2as}; q_1, q_2) G_3(e^{-4a} e^{-2as}; q_1, q_2, q_3)}{G_3(e^{-4a/\alpha}; q_1, q_2, q_3)} \tag{6.13}$$

(by immediately writing out the products on the r.h.s. of (6.13)). By the third equation in (6.8), $G_3(e^{-2as}; q_1, q_2, q_3) = G_2(e^{-2as}; q_1, q_2) G_3(e^{-4a} e^{-2as}; q_1, q_2, q_3)$, which means that we can obtain by (6.13) a new expression for $Z(s; \alpha, a, b)$:

$$Z(s; \alpha, a, b) = \frac{G_3(e^{-2as}; q_1, q_2, q_3)}{G_3(e^{-4a/\alpha} e^{-2as}; q_1, q_2, q_3)} \overset{i.e.}{=} \tag{6.14}$$

$$\Pi_{0 \leq k_1, k_2, k_2 \in \mathbb{Z}} \frac{[1 - e^{-4ak_3} (e^{i2b/\alpha})^{k_1} (e^{-2ib/\alpha})^{k_2} e^{-(k_1 + k_2 + \alpha s) 2a/\alpha}]}{[1 - e^{-2(k_3 + \frac{1}{\alpha}) 2a} (e^{i2b/\alpha})^{k_1} (e^{-i2b/\alpha})^{k_2} e^{-(k_1 + k_2 + \alpha s) 2a/\alpha}]}.$$

On the other hand, $q_1 q_2 \overset{def.}{=} e^{-4a/\alpha}$. Therefore by equation (6.12) we can use the first expression in (6.14) to also write

$$\begin{aligned} Z(s; \alpha, a, b) &= G_2(e^{-2as}; q_1, q_3) G_2(q_2 e^{-2as}; q_2, q_3) \\ &\overset{def.}{=} \Pi_{0 \leq k_1, k_2 \in \mathbb{Z}} [1 - (e^{-2a/\alpha} e^{2bi/\alpha})^{k_1} (e^{-4a})^{k_2} e^{-2as}] \cdot \\ \Pi_{0 \leq k_1, k_2 \in \mathbb{Z}} &\quad [1 - (e^{-2a/\alpha} e^{-2bi/\alpha})^{k_1} (e^{-4a})^{k_2} e^{-2a/\alpha} e^{-2bi/\alpha} e^{-2as}], \end{aligned} \tag{6.15}$$

which is precisely equation (4.2) as desired.

References

[1] M. Bañados, C. Teitelboim, J. Zanelli, Black hole in three-dimensional spacetime, Phys. Rev. Letters 69, no. 13, 1849-1851 (1992).

[2] A. Bytsenko, L. Vanzo, S. Zerbini, Quantum correction to the entropy of the (2+1)-dimensional black hole, Phys. Rev D 57, no. 8, 4917-4924 (1998).

[3] A. Bytsenko, M. Guimarães, F. Williams, Remarks on the spectrum and truncated heat kernel of the BTZ black hole, Letters in Math. Physics 79, 203-211 (2007).

[4] S. Carlip, C. Teitelboim, Aspects of black hole quantum mechanics and thermodynamics in 2+1 dimensions, Phys. Rev. D 51, no. 2, 622-631 (1995).

[5] J. Dodziuk, J. Jorgenson, Spectral asymptotics on degenerating hyperbolic 3 manifolds, Memoirs of the Am. Math. Soc. 135, no. 643 (1998).

[6] R. Gangolli, Zeta functions of Selberg's type for compact space forms of symmetric spaces of rank one, Illinois J. Math. 21, 1-42 (1977).

[7] I. Gradshteyn, I. Ryzhik, Table of Integrals, Series, and Products, Corrected and Enlarged Edition, Academic Press (1980).

[8] I. Ichinose, Y. Satoh, Entropies of scalar fields on three-dimensional black holes, Nuclear Physics B 447, 340-370 (1995).

[9] N. Mandouvalos, Spectral theory and Eisenstein series for Kleinian groups, Proc. London Math. Soc. 57, 209-238 (1988).

[10] R. Mann, S. Solodukhin, Quantum scalar field on a three-dimensional (BTZ) black hole instanton: heat kernels, effective action, and thermodynamics, Phys. Rev. D, no. 6, 3622-3632 (1997).
[11] S. Patterson, The Selberg zeta-function of a Kleinian group - from Number Theory, Trace Formulas, and Discrete Groups: Symposium in Honor of Alte Selberg, Academic Press (1989).
[12] P. Perry, Trace formula for an abelian cyclic group in hyperbolic three-space, six page fax sent to F. Williams, unpublished (2001).
[13] P. Perry, Heat trace and zeta function for the hyperbolic cylinder in three dimensions, four page fax based in part on notes of F. Williams, unpublished (2001).
[14] P. Perry, F. Williams, Selberg zeta function for the BTZ black hole, Internat. J. of Pure and Applied Math. 9, 1-21 (2003).
[15] F. Williams, A zeta function for the BTZ black hole, Internat. J. Modern Physics A18, 2205-2209 (2003).
[16] F. Williams, BTZ black hole and Jacobi inversion for fundamental domains of infinite volume, Contemporary Math. 376, 385-391 (2005).
[17] F. Williams, Conical defect zeta function for the BTZ black hole, Proceedings of the Einstein Symposium (One hundred years of Relativity 1905-2005). Scientific Annals of Alexandru Ioan Cuza Univ., Iais, Romania, 54-58 (2005).
[18] F. Williams, Note on quantum correction to BTZ instanton entropy, Proceedings of Science, Electronic Journal: POS (IC 2006) 006.
[19] F. Williams, A deformation of the Patterson-Selberg zeta function, to appear in Proceedings of the XVI Coloquio LatinoAmericano De Álgebra, Colonia, Uruguay, 2005.

DEPARTMENT OF MATHEMATICS AND STATISTICS, UNIVERSITY OF MASSACHUSETTS, AMHERST, MASSACHUSETTS 01003

E-mail address: `williams@math.umass.edu`

Contemporary Mathematics
Volume **467**, 2008

On the geometry of coisotropic submanifolds of Poisson manifolds

Aïssa Wade

ABSTRACT. We review basic results about Poisson manifolds and their coisotropic submanifolds. We show that the classical existence theorem for coisotropic embeddings of presymplectic manifolds can be extended to the case of Dirac manifolds. We give local normal forms describing these coisotropic embeddings.

1. Introduction

Coisotropic submanifolds of symplectic manifolds are important tools for the study of mechanical systems with symmetries (see for instance [**FHST89**] and references therein). In the mid-eighties, Libermann and Marle [**LM87**] extended the notion of a coisotropic submanifold to the general setting of Poisson structures. The interest in such objects has been growing due to the fact that they are closely related to various physical theories. For instance, coisotropic submanifolds of Poisson manifolds represent boundary conditions in various topological field theories (see [**CF04**]). They provide a suitable framework for solving the quantization problem of mechanical systems [**BW97, B00, B05, BGHHW, CF04, CF07**]. Basically, the quantization problem is the construction of a quantum theory for a given classical mechanical system. In the physics literature, coisotropic submanifolds appear under the name of first-class constraints. Here, we discuss only their geometric aspects.

It is known that any presymplectic manifold whose presymplectic form has constant rank can be coisotropically embedded into some symplectic manifold that consists of a tubular neighborhood of the zero section of the dual of its characteristic bundle together with an adapted symplectic structure [**G84**]. This result can be extended to the general setting of Dirac manifolds as observed by Cattaneo and Zambon [**CZ06**]. Here, we present a new proof of this result. We also give explicit expressions for coisotropic embeddings by using local normal forms of Dirac structures given in [**DW04, V06**].

1991 *Mathematics Subject Classification.* Primary: 53Dxx, 58xx.
Key words and phrases. Poisson structures, coisotropic submanifolds, Dirac structures.

The article is organized as follows: In Sections 2 and 3, we review basic concepts and known facts about Poisson manifolds, coisotropic submanifolds of Poisson manifolds and Dirac manifolds. In Section 4, we discuss the existence and the uniqueness of coisotropic embeddings of Dirac manifolds and provide an example. In Section 5, we give local normal forms describing coisotropic embeddings of Dirac manifolds and conclude with some remarks.

Acknowledgment. I would like to thank the organizers of CAARMS13, Donald King, William Massey, and Alfred Noël for the excellent organization of the meeting.

2. Preliminaries

2.1. Poisson structures.

DEFINITION 2.1. Let M be a smooth finite-dimensional manifold. A *Poisson structure* on M is an $\mathbb{R}$-bilinear antisymmetric operation $\{\ ,\ \}$ on the space $C^\infty(M)$ of smooth real-valued functions on M satisfying the following identities:

$$\text{(1)} \qquad \textit{Jacobi identity:} \qquad \{\{f,g\},h\}+\{\{g,h\},f\}+\{\{h,f\},g\}=0$$

$$\text{(2)} \qquad \textit{Leibniz identity:} \qquad \{f,gh\}=\{f,g\}h+g\{f,h\},$$

for all $f,g,h\in C^\infty(M)$.

In other words, the space $C^\infty(M)$ is equipped with a Lie bracket $\{\ ,\ \}$ which, additionally, satisfies the Leibniz identity. The operation $\{\ ,\ \}$ is also called a *Poisson bracket.* A manifold M endowed with a Poisson structure is called a *Poisson manifold.*

EXAMPLE 2.1. Every manifold admits a trivial Poisson structure given by $\{f,g\}=0$ for all functions f and g.

EXAMPLE 2.2. Consider $M=\mathbb{R}^{2n}$ with its standard coordinates (x_i,y_i) and define a Poisson structure on it by setting

$$\{f,g\}=\sum_{i=1}^{n}\Big(\frac{\partial f}{\partial x_i}\frac{\partial g}{\partial y_i}-\frac{\partial f}{\partial y_i}\frac{\partial g}{\partial x_i}\Big), \tag{3}$$

for all $f,g\in C^\infty(M)$. One can easily check that both the Jacobi identity and the Leibniz identity are satisfied for this bracket.

EXAMPLE 2.3. By a *non-degenerate* Poisson structure on a $2n$-dimensional manifold M, we mean a bilinear operation which is related to a closed 2-form Ω of rank $2n$ as follows:

$$\{f,g\}=\Omega(X_f,X_g), \tag{4}$$

where X_f is the Hamiltonian vector field given by

$$i_{X_f}\Omega=-df,\quad \forall\ f\in C^\infty(M).$$

The Leibniz identity follows from $d(gh)=gdh+hdg$. While the Jacobi identity for the bracket (4) is equivalent to the fact $d\Omega=0$. In fact, Ω is called a *symplectic form.* Non-degenerate Poisson structures are exactly symplectic structures.

EXAMPLE 2.4. The dual $\mathfrak{g}^*$ of any finite-dimensional Lie algebra $(\mathfrak{g}, [\ ,\ ])$ admits a canonical Poisson structure, called *a Lie-Poisson* structure and defined by:

$$\{f,g\}(\alpha) = \langle \alpha, [df(\alpha), dg(\alpha)] \rangle, \quad \forall\, f,g \in C^\infty(\mathfrak{g}^*) \quad \forall \alpha \in \mathfrak{g}^*, \tag{5}$$

where $T_\alpha \mathfrak{g}^* \simeq \mathfrak{g}^*$, the terms $df(\alpha)$ and $dg(\alpha)$ are consider as elements of the Lie algebra $\mathfrak{g}$.

EXAMPLE 2.5. **Product of Poisson manifolds**. Let $(M_1, \{\ ,\ \}_1)$ and $(M_2, \{\ ,\ \}_2)$ be two Poisson manifolds. On the space of smooth real-valued functions on $M_1 \times M_2$, we define the following bracket:

$$\{f,g\}(x_1,x_2) = \{f_{x_2}, g_{x_2}\}_1(x_1) + \{f_{x_1}, g_{x_1}\}_2(x_2), \tag{6}$$

where

$$f_{x_1}(x_2) = f_{x_2}(x_1) = f(x_1,x_2),$$

for any $f \in C^\infty(M_1 \times M_2)$, $x_1 \in M_1$ and $x_2 \in M_2$. One can easily check that the above bracket is a Poisson bracket on $M_1 \times M_2$. It is called a *product Poisson structure.*

REMARK 2.1. Every Poisson structure $\{\ ,\ \}$ on M is uniquely characterized by an associated bivector field π defined as follows:

$$\pi(df, dg) = \{f,g\}, \qquad f,g \in C^\infty(M).$$

Conversely, a bivector field π on M defines a Poisson structure if and only if the Schouten bracket $[\pi,\pi]$ is zero (see [**V94, DZ05**]). In local coordinates $(x_1, \dots, x_d)$, one has the expression:

$$\pi = \sum_{i<j} \pi_{ij} \frac{\partial}{\partial x_i} \wedge \frac{\partial}{\partial x_j},$$

where the π_{ij} are smooth functions. The condition $[\pi,\pi] = 0$ means that

$$\sum_{s=1}^{d} \left(\frac{\partial \pi_{ij}}{\partial x_s} \pi_{sk} + \frac{\partial \pi_{jk}}{\partial x_s} \pi_{si} + \frac{\partial \pi_{ki}}{\partial x_s} \pi_{sj} \right) = 0. \tag{7}$$

Such a bivector field π is called a *Poisson tensor* on M. It determines a bundle morphism $\pi^\sharp : T^*M \to TM$ given by:

$$\langle \beta, \pi^\sharp \alpha \rangle = \pi(\alpha, \beta),$$

for all $\alpha, \beta \in T^*M$.

2.2. Coisotropic submanifolds. Let C be a submanifold of a Poisson manifold $(M, \{\ ,\ \})$. Consider the *vanishing ideal* $I_C = \{f \in C^\infty(M) \mid f_{|C} = 0\}$.

DEFINITION 2.2. The submanifold C is *coisotropic* if I_C is closed under the Poisson bracket $\{\ ,\ \}$.

Now we review some known facts about coisotropic submanifolds.

a. Let π be the corresponding Poisson tensor on M and let N^*C be the conormal bundle of C, i.e.

$$N^*C(x) = \{\alpha_x \in T_x^*M \mid \langle \alpha_x, u \rangle = 0, \forall\, u \in T_xC\}.$$

Then C is a coisotropic submanifold of M if and only if

$$\pi^\sharp(N^*C) \subset TC.$$

b. Now suppose π defines a non-degenerate Poisson structure on M and let Ω be its corresponding symplectic form. Then the map $\pi^\sharp : N^*C \to \mathrm{Orth}_\Omega TC$ is an isomorphism, where $\mathrm{Orth}_\Omega TC$ is given by:

$$\mathrm{Orth}_\Omega TC = \{v \in TM \mid \Omega(u, v) = 0, \ \forall u \in TC\}.$$

In the non-degenerate case, C is coisotropic if and only if

$$\mathrm{Orth}_\Omega TC \subset TC.$$

c. There are coisotropic submanifolds of Poisson manifolds M whose intersections with symplectic leaves are not coisotropic. Under some transversality, any intersection with a symplectic leaf is coisotropic. More precisely, assume that C intersects cleanly the symplectic leaves of M, i.e.

$$C \cap F \text{ is a manifold and } T(C \cap F) = TC \cap TF,$$

for every symplectic leaf F of M, then $C \cap F$ is coisotropic. We should mention that reductions of coisotropic submanifolds were studied in Poisson geometry [**MR86**].

2.3. Coisotropic embeddings of presymplectic manifolds. Recall that a *presymplectic form* on a smooth manifold M is a closed 2-form (not necessarily non-degenerate). A manifold equipped with a presymplectic form is called a *presymplectic manifold.* When the rank of the presymplectic form ω is constant then its kernel $\mathrm{Ker}\omega$ defines an involutive distribution. It determines a foliation called a *characteristic foliation.* We have the following coisotropic embedding result:

PROPOSITION 2.1. [**G84**] *Given any presymplectic manifold (C, ω) whose presymplectic form has constant rank, there exists a symplectic form on a tubular neighborhood of the zero section of the dual bundle K^* of $K = \mathrm{Ker}\omega$ such that C can be coisotropically embedded into this neighborhood.*

Moreover, any two coisotropic embeddings of (C, ω) are locally equivalent (up to symplectomorphism).

A proof of this proposition, which uses methods in [**We81**], can be found in [**G84**]. Another proof is given in [**OP05**].

3. Dirac manifolds

3.1. Basic definitions and examples. The *Courant bracket* [**C90**] on the space of smooth sections of the vector bundle $TM \oplus T^*M \to M$ is a natural extension of the Lie bracket of vector fields on M. Precisely, it is defined by:

$$[(X_1, \alpha_1), \ (X_2, \alpha_2)] = \Big([X_1, X_2], \ \mathcal{L}_{X_1}\alpha_2 - i_{X_2}d\alpha_1\Big), \tag{8}$$

for all (X_1, α_1) and $(X_2, \alpha_2) \in \Gamma(TM \oplus T^*M)$, where $[X_1, X_2]$ is the usual Lie bracket of vector fields, $\mathcal{L}_{X_1}\alpha_2$ the Lie derivative of the 1-form α_2 along X_1.

DEFINITION 3.1. [**C90**] A *Dirac structure* on a smooth d-dimensional manifold M is a rank d subbundle L of $TM \oplus T^*M \to M$ whose space of smooth sections is closed under the Courant bracket and such that:

$$\langle\langle (X_1, \alpha_1), \ (X_2, \alpha_2) \rangle\rangle = \frac{1}{2}\Big(\alpha_1(X_2) + \alpha_2(X_1)\Big) = 0, \tag{9}$$

for all smooth sections (X_1, α_1) and $(X_2, \alpha_2) \in \Gamma(TM \oplus T^*M)$. A manifold equipped with a Dirac structure is called a *Dirac manifold.*

EXAMPLE 3.1. [**C90**] Presymplectic structures are in one-to-one correspondence with Dirac structures L with $L \cap (\{0\} \times T^*M) = \{0\}$. Indeed, given any presymplectic form Ω, its graph

$$L_\Omega = \{(X, i_X\Omega) \mid X \in TM\}$$

defines a Dirac structure on M with $L \cap (\{0\} \times T^*M) = \{0\}$. Conversely, any Dirac structure on M satisfying this condition determines a presymplectic form on M.

EXAMPLE 3.2. There is a one-to-one correspondence between Poisson structures and Dirac structures L on M satisfying $L \cap (TM \times \{0\}) = \{0\}$. The Dirac structure associated with a Poisson structure π is:

$$L_\pi = \{(\pi^\sharp\alpha, \alpha) \mid \alpha \in T^*M\}.$$

3.2. Presymplectic foliation of a Dirac manifold. Let (M, L) be a Dirac manifold and $\mathrm{pr} : L \to TM$ the natural projection of L onto TM. Then $\mathcal{D} = \mathrm{pr}(L)$ is an involutive distribution which determines a foliation, i.e. a partition of M into connected and immersed submanifolds, called leaves. The dimensions of the leaves may vary. But each leaf of M carries a presymplectic form defined at the point $x \in M$ as follows:

$$\theta_x(X_1, X_2) = \alpha_1(X_2), \tag{10}$$

where α_1 is any element in T_x^*M satisfying $(X_1, \alpha_1) \in L_x$ and $(X_2, \alpha_2) \in L_x$. Equation (9) implies that (10) is independent of the choice of the covector α_1 satisfying $(X_1, \alpha_1) \in L_x$. The 2-forms θ_x fit together into a smooth leafwise 2-form θ. Using the fact that $\Gamma(L)$ is closed under the Courant bracket, one sees that θ is closed on each leaf.

3.3. Gauge transformations and Dirac maps. Now we will describe some methods which allow to construct new Dirac structures starting from a given one.

Gauge transformations. Let L be a Dirac structure on M and B a closed 2-form, then

$$\tau_B(L) := \{(X, \alpha + i_X B) \mid (X, \alpha) \in L\} \tag{11}$$

defines a Dirac structure on M [**SW01**]. The operation τ_B is called a *gauge transformation associated with B*. Basically, this operation modifies L by adding the pull-back of B to the presymplectic form on each leaf.

Dirac maps. Let (M_1, L_1) and (M_2, L_2) be two Dirac manifolds. A smooth map $f : M_1 \to M_2$ is called a *forward Dirac map* if L_1 and L_2 are related as follows:

$$Ł_2 = \{(df(Y),\ \alpha) \mid Y \in TM_1, \alpha \in T^*M_2 \text{ and } (Y,\ df^*(\alpha)) \in L_1\}. \tag{12}$$

We simplicity, we will write $L_2 = f_*L_1$.

A smooth map $g : M_1 \to M_2$ is called a *backward Dirac map* if

$$L_1 = \{(Y,\ dg^*(\alpha)) \mid Y \in TM_1, \alpha \in T^*M_2 \text{ and } (dg(Y),\ \alpha) \in L_2\}. \tag{13}$$

In this case, we write $L_1 = g^*L_2$.

Observe that g^*L_2 is a well-defined subbundle of TM_1 but it is not necessarily smooth. While f_*L_1 may not be well-defined. However, when g is a submersion, g^*L_2 is always a smooth Dirac structure.

EXAMPLE 3.3. Suppose M_1 and M_1 are Poisson manifolds. Then $f : M_1 \to M_2$ is a forward Dirac map if and only if its graph $Gr(f)$ is a coisotropic submanifold of the product Poisson manifold $M_1 \times M_2$.

EXAMPLE 3.4. Let L be a Dirac structure on M. We consider a presymplectic leaf F of M, equipped with a Dirac structure L_ω associated with the presymplectic form ω. The inclusion map $\iota : F \hookrightarrow M$ is both a backward map and a forward map [**BR03**].

4. Coisotropic embeddings of Dirac manifolds

4.1. Existence Theorem. In this section, we are interested in the following question: Given a Dirac manifold C, does there exist a coisotropic embedding of (C, L) into some adapted Poisson manifold? First all, if $K = L \cap (TC \times \{0\}) = \{0\}$ then C is a Poisson manifold. In this case, there is nothing to prove. Now suppose that K is non-trivial. We aim to generalize Gotay's embedding theorem for presymplectic manifolds to the setting of Dirac structures. We obtain:

THEOREM 4.1. *Let C be a smooth finite-dimensional manifold endowed with a Dirac structure L such that the vector bundle $K = L \cap (TC \times \{0\})$ has constant rank. Then, there exist a neighborhood U of the zero section in the dual K^* of K and a Poisson structure π_U on U such that C can be coisotropically embedded into (U, π_U).*

PROOF. Since K has constant rank, it determines a foliation $\mathcal{K}$ on C. Consider the zero section embedding $\iota : C \hookrightarrow K^*$ and identify K with its first projection onto TC. Let $\Gamma : TC \to K$ be a bundle map such that $\Gamma^2 = \Gamma$. Then $TC = \mathrm{Ker}(\Gamma) \oplus K$. The map Γ can be considered as a K-valued 1-form, which is preserved by the flow generated by any vector field X tangent to the foliation $\mathcal{K}$, i.e. $\mathcal{L}_X\Gamma = 0$ (see details about the local expressions of Γ in Section 5). There is a corresponding 1-form θ_Γ on K^* defined as follows:

$$\theta_\Gamma(\alpha)(Z) = \langle \alpha, \Gamma \circ dp(Z) \rangle, \tag{14}$$

where $p : K^* \to C$ is the canonical projection of K^* onto C, $\alpha \in K^*$, and $Z \in T_\alpha K^*$.

Furthermore, there is an induced foliation on K^* whose leaves are the connected components of the preimages under p of the leaves of $\mathcal{K}$. Let $\mathrm{Vert} \subset TK^*$ be the tangent distribution to that induced foliation. On a small tubular neighborhood U of C in K^*, the 2-form θ_Γ is non-degenerate on the vertical subspaces Vert_α (see the local expressions in Section 5). Moreover, it determines a field of horizontal subspaces:

$$\mathrm{Hor}_\Gamma(\alpha) = \{Y_\alpha \in T_\alpha K^* \mid d\theta_\Gamma(X, Y) = 0, \ \forall\, X \in \mathrm{Vert}_\alpha\}.$$

Choose a real number $\lambda > 0$ and define the Dirac structure L_λ on K^* by setting

$$L_\lambda = \tau_{d(\lambda\theta_\Gamma)}(p^* L),$$

where $\tau_{d(\lambda\theta_\Gamma)}$ is the gauge transformation associated with $d(\lambda\theta_\Gamma)$. Then, L_λ determines a Dirac structure which is a Poisson structure on K^* provided that λ is small enough. There follows the result.

□

REMARK 4.1. Cattaneo and Zambon [**CZ06**] obtained a result similar to Theorem 4.1. Their description is different from the one given in the above proof and it is implicit.

A priori, the Poisson structure corresponding to L_λ depends on the choices of Γ and λ. Suppose that $L \cap (TC \times \{0\})$ defines a regular and *simple* foliation $\mathcal{K}$. In this case, L is said to be *reducible.* It induces a Poisson structure on the quotient manifold $C/\mathcal{K}$, which is Hausdorff and smooth. Under the additional assumption that L is reducible, one can show all coisotropic embeddings of (C, L) are locally equivalent (up to Dirac maps). The proof, which is quite technical, will not be presented here.

4.2. Example. Consider $C = S^3 \times \mathfrak{so}^*(3)$, i.e. the product of the sphere $S^3 \subset \mathbb{R}^4$ with the dual of the Lie algebra $\mathfrak{so}(3)$. Here, $\mathfrak{so}^*(3)$ is endowed with its canonical Lie-Poisson structure Λ, which is inherited from the Lie algebra structure on $\mathfrak{so}(3)$, while S^3 is equipped with its contact form $\alpha = i^*\sigma$, where $\sigma = 1/2 \sum (p_i dq_i - q_i dp_i)$ is the Liouville form on $\mathbb{R}^4$ and $i : S^3 \hookrightarrow \mathbb{R}^4$ is the inclusion map. Consider the Dirac structure L on C whose space of sections is spanned by the sections of the form

$$(X,\ i_X d\alpha), \quad (\Lambda^\sharp \beta,\ \beta),$$

where $X \in \mathfrak{X}(S^3)$ and $\beta \in \Omega^1(\mathfrak{so}^*(3))$. Clearly, $(S^3 \times \mathfrak{so}^*(3), L)$ is a reducible Dirac manifold. Its characteristic foliation is determined by the orbits of the flow of the Reeb vector field $\mathbf{Z}$ on S^3 , i.e. the circles of the Hopf fibration. One can see that $(S^3 \times \mathfrak{so}^*(3), L)$ embeds coisotropically into $M = \mathbb{R} \times S^3 \times \mathfrak{so}^*(3)$ with the product Poisson structure $\omega^{-1} \oplus \Lambda$, where $\omega = d(e^t \alpha)$ is the symplectization of the contact structure on S^3.

5. Local normal forms

In this section, we give the local expressions of the above Dirac structure L_λ.

PROPOSITION 5.1. [**DW04, V06**] *Let L be a Dirac structure on C such that $K = L \cap (TC \oplus \{0\})$ has constant rank. Given any point $q \in C$, there are coordinates $(x_1, \ldots, x_r, x'_1, \ldots, x'_s, y_1, \ldots, y_n)$ defined on an open neighborhood W of q in C such that $L_{|W}$ is spanned by sections of the form*

$$\begin{aligned}
F_j &= \left(\frac{\partial}{\partial y_j},\ 0\right) \\
G_i &= \left(\frac{\partial}{\partial x_i} + \sum_j R_i^j(x,x')\frac{\partial}{\partial x'_j},\ \sum_j \omega_i^j(x,x')dx_j\right) \\
G'_k &= \left(\sum_\ell \pi_{k\ell}(x,x')\frac{\partial}{\partial x'_\ell},\ dx'_k - \sum_l R_\ell^k(x,x')dx_l\right),
\end{aligned}$$

where the $R_i^j(x,x')$ and $\pi_{k\ell}(x,x')$ vanish at $(x,x') = (0,0)$.

Proposition 5.1 is a consequence of the local normal forms given in [**DW04**]. A complete proof of this result can be found in [**V06**]. In fact, this proposition ensures the existence of a local coordinate system $(x,y) = (x_1, \ldots, x_m, y_1, \ldots, y_n)$ defined on a neighborhood W of any point q in C such the intersections of W with the leaves of $\mathcal{K}$ are given by the equations

$$x_1 = c_1,\ \ldots,\ x_m = c_m, \quad \text{with } c_i \in \mathbb{R},$$

and $L_{|W}$ is determined by sections of the form

$$F_j = \Big(\frac{\partial}{\partial y_j}, 0\Big) \quad \text{and} \quad E = \Big(\sum_{i=1}^{m} h_i \frac{\partial}{\partial x_i}(x),\ \sum_{i=1}^{m} \alpha_i(x) dx_i\Big),$$

where the functions h_i and α_i depend only on the x-coordinates. We have the local expressions:

$$\Gamma = \sum_{i=1}^{n} \Big(dy_i + \Gamma_i^j dx_j\Big) \otimes \frac{\partial}{\partial y_i}. \tag{15}$$

By a simple computation, one gets

$$\theta_\Gamma = \sum_{i=1}^{n} z_i (dy_i + \Gamma_i^j dx_j),$$

where $(x_1, \ldots, x_m, y_1, \ldots, y_n, z_1, \ldots, z_n)$ are canonical coordinates defined on an open neighborhood of V_q of q in K^* such that the z_i's are linear on the fibers of K^*. One gets

$$d\theta_\Gamma \quad = \quad \sum_{i,j} dz_i \wedge (dy_i + \Gamma_i^j dx_j) + \sum_{i,j,\ell} z_i \frac{\partial \Gamma_i^j}{\partial x_\ell} dx_\ell \wedge dx_j + \sum_{i,j,k} z_i \frac{\partial \Gamma_i^j}{\partial y_k} dy_k \wedge dx_j.$$

Since $\mathcal{L}_X \Gamma = 0$, for any vector field X tangent to the foliation $\mathcal{K}$, it follows

$$d\theta_\Gamma = \sum_{i,j,k,u} dz_i \wedge (dy_i + \Gamma_i^j dx_j) + \sum_{i,j,\ell} z_i \frac{\partial \Gamma_i^j}{\partial x_\ell} dx_\ell \wedge dx_j.$$

Now, we consider local smooth sections of L having the expression:

$$h = \Big(\sum_i h_i(x) \frac{\partial}{\partial x_i},\ \sum_i \alpha_i(x) dx_i\Big) \quad \text{with} \quad (\alpha_1, \ldots, \alpha_m) \neq (0, \ldots, 0).$$

We want to find all vector fields of the form

$$Y = \sum_i h_i(x) \frac{\partial}{\partial x_i} + \sum_j \mu_j(x, y, z) \frac{\partial}{\partial y_j} + \sum_k \nu_k(x, y, z) \frac{\partial}{\partial z_k}$$

satisfying the equations:

$$d\theta_\Gamma\Big(Y, \frac{\partial}{\partial y_k}\Big) = d\theta_\Gamma\Big(Y, \frac{\partial}{\partial z_k}\Big) = 0,$$

for all $k \in \{1, \ldots, n\}$. Solving these equations one gets

$$Y = \sum h_i \left(\frac{\partial}{\partial x_i} - \sum_j \Gamma_j^i \frac{\partial}{\partial y_j}\right).$$

We set

$$\mathbf{X}_i = \frac{\partial}{\partial x_i} - \sum_j \Gamma_j^i \frac{\partial}{\partial y_j}.$$

We have obtained the following proposition:

PROPOSITION 5.2. *On the open neighborhood V_q, the space of local sections of L_λ is spanned by elements of the form*

$$\begin{aligned} e_i &= \Big(\frac{\partial}{\partial z_i},\ dy_i + \sum_j \Gamma_i^j dx_j\Big), \\ e_i' &= \Big(\frac{\partial}{\partial y_i},\ -dz_i\Big) \\ e'' &= \Big(\sum_i h_i(x)\mathbf{X}_i,\ \sum_i (\alpha_i(x) + \lambda\beta_i)dx_i\Big), \end{aligned}$$

where the β_i are components of $i_Y d\theta_\Gamma$.

These local expressions of L_λ show that it is a maximally isotropic subbundle with respect to the symmetric operation $\langle\ ,\ \rangle$, i.e. Equation (9) is satisfied. When λ is small enough, $\alpha_i + \lambda\beta_i \neq 0$. Moreover, one gets

$$L_\lambda \cap (TV_q \oplus \{0\}) = \{0\}. \tag{16}$$

Hence L_λ determines a Poisson structure on V_q.

□

REMARK 5.1. **a.** Local expressions for coisotropic embeddings of presympletic manifolds were given in [**OP05**]. Our local normal forms generalize their formulas.

b. As announced above (Remark 4.1), the uniqueness of coisotropic embeddings of Dirac manifolds, up to local Dirac maps that coincide with the identity on C, can be established when L is a reducible Dirac structure on C. Another particular case where the leafwise presymplectic form on C has constant rank is treated in [**CZ06**]. However, the general case is still an open problem.

References

[BW97] S. Bates and A. Weinstein, *Lectures on the Geometry of Quantization*, Berkeley Mathematics Lecture Notes **8**, AMS, Providence, RI, 1997.

[B00] M. Bordemann, *The deformation quantization of certain super-Poisson brackets and BRST cohomology*, Conference Moshé Flato, Math. Phys. Stud., **22**, Kluwer Acad. Publ., Dordrecht, 2000, 45–68.

[B05] M. Bordemann, *(Bi)modules, morphisms, and reduction of star-products: the symplectic case, foliations, and obstructions*, Travaux mathmatiques, Fasc. **XVI**, Luxembourg, 2005, 9–40.

[BGHHW] M. Bordemann, G. Ginot, G. Halbout, H. Hans-Christian, and S. Waldmann, *Formalité G_∞ adaptée et star-representations des sous-variétés coisotropes*, Preprint arXiv:math/0504276.

[BR03] H. Bursztyn and O. Radko, *Gauge equivalence of Dirac structures and symplectic groupoids*, Ann. Inst. Fourier (Grenoble) **53** (2003), 309–337.

[CF04] A. Cattaneo and G. Felder, *Coisotropic submanifolds in Poisson geometry and branes in the Poisson sigma model*, Lett. Math. Phys. **69** (2004), 157–175.

[CF07] A. Cattaneo and G. Felder, *Relative formality theorem and quantisation of coisotropic submanifolds*, Adv. Math. **208** (2007), 521–548.

[CZ06] A. Cattaneo and M. Zambon, *Coisotropic embdeddings in Poisson manifolds*, Preprint Math.SG/0611480.

[C90] T. Courant, *Dirac structures*, Trans. A.M.S. **319** (1990), 631-661.

[DW04] J.-P. Dufour and A. Wade, *On the local structure of Dirac manifolds*, arXiv:math/0405257, to appear in Compositio Mathematica.

[DZ05] J.-P. Dufour and N.-T. Zung, *Poisson structures and their normal forms*, Progress in Mathematics **242**, Birkhäuser Verlag, Basel, 2005.

[FHST89] J. Fisch, M. Henneaux, J. Stasheff, and C. Teitelboim, *Existence, uniqueness and cohomology of the classical BRST charge with ghosts of ghosts*, Comm. Math. Phys. **120** (1989), 379–407.

[G84] M.-J. Gotay, *On coisotropic embeddings of presymplectic manifolds*, Proc. Amer. Math. Soc. **84** (1982) 111–114.

[LM87] P. Libermann and C.-M. Marle, *Symplectic geometry and analytical mechanics*, Mathematics and its Applications, **35**. Reidel Publishing Co., Dordrecht, 1987.

[MR86] J. Marsden and T. Ratiu, *Reduction of Poisson manifolds*, Lett. Math. Phys. **11** (1986), 161–169.

[OP05] Y.-G. Oh and J.-S. Park, *Deformations of coisotropic submanifolds and strong homotopy Lie algebroids*, Invent. Math. **161** (2005), 287–360

[SW01] P. Severa and A. Weinstein, *Poisson geometry with a 3-form background*, Progr. Theoret. Phys. Suppl. No. **144** (2001), 145–154

[V94] I. Vaisman, *Lectures on the geometry of Poisson manifolds*, Progress in Mathematics, **118**. Birkhäuser Verlag, Basel, 1994.

[V06] I. Vaisman, *Foliation-coupling Dirac structures*, J. Geom. Phys. **56** (2006), 917–938.

[We81] A. Weinstein, *Neighborhood classification of isotropic embeddings*, J. Differential Geom. **16** (1981), no. 1, 125–128.

[We77] A. Weinstein, *Lectures on symplectic manifolds*, Regional Conference Series in Mathematics **29**, American Mathematical Society, Providence, R.I., 1977.

Aissa Wade, Department of Mathematics, Penn State University, PA 16802.
E-mail address: wade@math.psu.edu

Contemporary Mathematics
Volume **467**, 2008

Remarks on Ultrafilters on the Collection of Finite Subsets of an Infinite Set

Arthur D. Grainger

Abstract. Let J be an infinite set and let $I = \mathcal{P}_f(J)$, i.e., I is the collection of all non empty finite subsets of J. Let βI denote the collection of all ultrafilters on the set I. In this survey article, we consider $(\beta I, \uplus)$, the compact (Hausdorff) right topological semigroup that is the *Stone – Čech Compactification* of the semigroup $(I, \cup)$ equipped with the discrete topology. It is shown that there is an injective map $A \to \beta_A(I)$ of $\mathcal{P}(J)$ into $\mathcal{P}(\beta I)$ such that each $\beta_A(I)$ is a closed subsemigroup of $(\beta I, \uplus)$, the set $\beta_J(I)$ is a closed ideal of $(\beta I, \uplus)$ and the collection $\{\beta_A(I) \mid A \in \mathcal{P}(J)\}$ is a partition of βI. The algebraic structure of βI is explored. In particular, it is shown that $\beta_J(I) = K(\beta I)$, i.e., $\beta_J(I)$ is the smallest ideal of βI.

1. Introduction

In this survey article, we review the recent results on the structure of the set of ultrafilters on the collection of finite subsets of an infinite set, as featured in [**1**], [**2**] and [**4**]. The semigroup of ultrafilters has been studied since it was introduced in the late 1950's. The collection of ultrafilters is intrinsically interesting as being an extension of the underlying set. If a set has a semigroup structure, then the collection of all ultrafilters on it is the largest semigroup compactification of the set.

1.1. Preliminaries. First things first, let's formally review what filters and ultrafilters are.

Definition 1. *A non empty collection p of subsets of a set S is a **filter** if and only if*

- $\varnothing \notin p$,
- $F \in p$ *and* $F \subseteq V$ *imply* $V \in p$,
- $F_1 \in p$ *and* $F_2 \in p$ *imply* $F_1 \cap F_2 \in p$.

*A filter p is called an **ultrafilter** if and only if*

- *for* $X \subseteq S$, *either* $X \in p$ *or* $S \backslash X \in p$.

1991 *Mathematics Subject Classification.*
Key words and phrases. ultrafilter, semigroup, smallest ideal .

In other words, an ultrafilter is a maximal filter; i.e., if p is an ultrafilter and $\mathcal{F}$ is a filter such that $p \subseteq \mathcal{F}$, then $p = \mathcal{F}$.

We will denote the collection of all ultrafilters on a set S by βS.

DEFINITION 2. *A collection $\mathcal{A}$ of subsets S has the **finite intersection property** (F.I.P.) if and only if $\{A_i\}_{i=1}^{n} \subseteq \mathcal{A}$ implies $\bigcap_{i=1}^{n} A_i \neq \varnothing$.*

By Zorn's lemma, we have that if a collection $\mathcal{A}$ has the finite intersection property, there is an ultrafilter p such that $\mathcal{A} \subseteq p$. Consequently, any filter is a sub-collection of some ultrafilter.

EXAMPLE 1 (of Filters and Ultrafilters). *Let S be a non empty set.*

- *Let τ be a topology on S. For $x \in S$, the collection $\mathcal{N}_\tau(x)$ of all $\tau-$neighborhoods of x is a filter (on S).*
- *Assume that S is infinite and let $\mathcal{F}_r(S)$ be the collection of all subsets of S that have finite complements. The collection $\mathcal{F}_r(S)$ is a filter [called the Fréchet filter on S].*
- *For $x \in S$, the collection $e(x) = \{B \subseteq S \mid x \in B\}$ is an ultrafilter and is called the **principal ultrafilter** generated by x. The map $x \to e(x)$ is an injection of S into βS. All other elements of βS are called **non principal ultrafilters** (or **free ultrafilters**).*

It can be shown that an ultrafilter $p \in \beta S$ is free if and only if $\bigcap p = \varnothing$. Also, for an infinite set S, p is free if and only if $\mathcal{F}_r(S) \subseteq p$.

A collection $\mathcal{A}$ of subsets S is a **filter basis** for a filter $\mathcal{F}$ (on S) if and only if $\mathcal{A} \subseteq \mathcal{F}$ and for $F \in \mathcal{F}$, there exists $A \in \mathcal{A}$ such that $A \subseteq F$. Such a collection $\mathcal{A}$ is said to **generate** the filter $\mathcal{F}$. For example, let $g : S \to Y$ be a mapping of S into some set Y and let $\mathcal{F}$ be a filter on S. The collection

$$\{\, g[F] \mid F \in \mathcal{F}\}$$

generates a filter on Y [denoted by $g[\mathcal{F}]$ and is called the image filter of $\mathcal{F}$ under g]. Also, if p is an ultrafilter on S, then $g[p]$ is an ultrafilter on Y.

Filters can be used to generalize the concept of convergence and ultrafilters are used to characterize compactness. Specifically, if τ is a topology on a set S, $\mathcal{F}$ is a filter on S and $x \in S$, then $\mathcal{F}$ is said to converge to x if and only if $\mathcal{N}_\tau(x) \subseteq \mathcal{F}$, where $\mathcal{N}_\tau(x)$ is the filter of all $\tau-$neighborhoods of x. It can be shown that a topological space (S, τ) is compact if and only if every ultrafilter on S converges (see [**3**], Theorem 3.52 on page 65).

DEFINITION 3. *Let S be a set. For $E \subseteq S$, define*

$$\widehat{E} = \{\, p \in \beta S \mid E \in p\}.$$

The collection $\left\{\, \widehat{E} \mid E \subseteq S \,\right\}$ generates a topology τ on βS that is the Stone-Čech compactification of (S, d), where d is the discrete topology on S (i.e., $d = \mathcal{P}(S)$). That is:

- The topology τ on βS generated by $\left\{\, \widehat{E} \mid E \subseteq S \,\right\}$ is compact and Hausdorff (T_2).
- $V \subseteq \beta S$ is clopen if and only if
$$V \in \left\{\, \widehat{E} \mid E \subseteq S \,\right\}.$$

- The collection $\{ e(x) \mid x \in S \}$ of all principal ultrafilter on S is dense in $(\beta S, \tau)$.
- Any map $f : S \to Y$ of S into a compact, T_2 space Y extends uniquely to a $\tau-$continuous map $\tilde{f} : \beta S \to Y$.

[see chapter 3 of [**3**]].

REMARK 1. *It can be shown that for* $E \subseteq S$, $\widehat{E}$ *is the* $\tau-$*closure of* $e[E] = \{ e(x) \mid x \in E \}$ *[see* [**3**], *Theorem 3.18 (c) on page 53].*

REMARK 2. *We will call the topology* τ *on* βS *[generated by* $\left\{ \widehat{E} \mid E \subseteq S \right\}$*] the Stone-Čech topology on* βS *[or the S-C topology on* βS *].*

The above brief review establishes the importance of filters and ultrafilters. Now, we will examine the structure of the space of all ultrafilters on the collection of non empty finite subsets of an infinite set.

1.2. The Setting. Let J be an infinite set and let $I = \mathcal{P}_f(J)$, i.e., I is the collection of all non empty finite subsets of J. Let βI denote the collection of all ultrafilters on the set I. We denote the set of finite ordinals by ω and the set of counting numbers by $\mathbb{N}$ [i.e., $\mathbb{N} = \omega \setminus \{0\}$]. We start with the following definitions from [**2**].

DEFINITION 4. *For* $A \subset J$, *define* $\mathcal{G}_A = \mathring{A} \cup \mathcal{B}_A$, *where* $\mathring{A} = \left\{ \mathring{j} \mid j \in A \right\}$, $\mathcal{B}_A = \left\{ I \setminus \mathring{j} \mid j \in J \setminus A \right\}$ *and* $\mathring{j} = \{ i \in I \mid j \in i \}$.

So, for $A \subseteq J$, we have

$$\mathcal{G}_A = \left\{ \mathring{j} \mid j \in A \right\} \cup \left\{ I \setminus \mathring{j} \mid j \in J \setminus A \right\}. \tag{1.1}$$

DEFINITION 5. *Define*

$$\mathcal{H} = \{ I \setminus [J]^n \mid n \in \mathbb{N} \},$$

where $[J]^n = \{ i \in I \mid |i| = n \}$ *and* $|i| \in \mathbb{N}$ *is the cardinality of* i.

For $A \subseteq J$, consider finite sets F, G and $\{n_t\}_{t=1}^{s}$ such that $F \subseteq A$, $G \subseteq J \setminus A$ and $\{n_t\}_{t=1}^{s} \subseteq \mathbb{N}$. If $i \in I$ such that $F \subseteq i$, $i \cap G = \varnothing$ and $\max(\{n_t\}_{t=1}^{s}) < |i|$, then

$$i \in \left[\bigcap_{j \in F} \mathring{j} \right] \cap \left[\bigcap_{j \in G} I \setminus \mathring{j} \right] \cap \left[\bigcap_{t=1}^{s} I \setminus [J]^{n_t} \right].$$

Consequently, the collection $\mathcal{G}_A \cup \mathcal{H}$ has the finite intersection property. In particular, collections $\mathcal{G}_A$ and $\mathcal{H}$ have the finite intersection property.

DEFINITION 6. *For* $A \subseteq J$, *define*

$$\beta_A(I) = \{ p \in \beta I \mid \mathcal{G}_A \subseteq p \} \qquad \textit{and}$$

$$\beta_0^{(A)}(I) = \{ p \in \beta I \mid \mathcal{G}_A \cup \mathcal{H} \subseteq p \}.$$

Clearly, $\beta_0^{(A)}(I) \subseteq \beta_A(I)$ and it can be shown that $\beta_0^{(A)}(I) = \beta_A(I)$ if and only if A is infinite [see [**2**], Propositions 4.2 and 4.3]. As noted in [**2**], The ultrafilters in $\beta_J(I)$ have played a significant role in other areas of mathematics, especially in nonstandard analysis. Specifically, elements of $\beta_J(I)$ are used to construct ultrapowers of a set that are enlargements (when viewed as a nonstandard model of the set).

2. Algebraic Properties of βI

2.1. A Partition of βI.

PROPOSITION 1. *The collection* $\{\beta_A(I) \mid A \in \mathcal{P}(J)\}$ *is a partition of* βI.

PROOF. For $p \in \beta I$, let $A = \{j \in J \mid \overset{\circ}{j} \in p\}$. Hence, $p \in \beta_A(I)$. For $A, B \in \mathcal{P}(J)$, with $A \neq B$, we may assume there is some $j_0 \in A \backslash B$. Therefore, for each $p \in \beta_A(I)$, we have $\overset{\circ}{j}_0 \in p$ and for each $p \in \beta_B(I)$, we have $I \backslash \overset{\circ}{j}_0 \in p$. We infer $\beta_A(I) \cap \beta_B(I) = \varnothing$. □

The next proposition give a characterization of the *hat set* of $\overset{\circ}{j}$ for $j \in J$.

PROPOSITION 2. *If* $j \in J$, *then*

$$\widehat{\overset{\circ}{j}} = \bigcup \{\beta_D(I) \mid D \in e(j)\},$$

where $e(j) = \{D \in \mathcal{P}(J) \mid j \in D\}$ *[i.e.,* $e(j)$ *is the principal ultrafilter on* J *generated by* j*].*

PROOF. Assume $p \in \widehat{\overset{\circ}{j}}$. Hence, $\overset{\circ}{j} \in p$ [by definition of $\widehat{\overset{\circ}{j}}$]. Now, $p \in \beta_D(I)$ for some unique $D \in \mathcal{P}(J)$ [Proposition 1]. Consequently, $j \in D$. Therefore, $D \in e(j)$, which implies $p \in \bigcup \{\beta_D(I) \mid D \in e(j)\}$.

Conversely, assume $p \in \bigcup \{\beta_D(I) \mid D \in e(j)\}$. Hence, $p \in \beta_D(I)$ for some $D \in e(j)$. Note that $D \in e(j)$ implies that $j \in D$. So, $j \in D$ and $p \in \beta_D(I)$ imply $\overset{\circ}{j} \in p$, which implies $p \in \widehat{\overset{\circ}{j}}$ [by definition of $\widehat{\overset{\circ}{j}}$].

We infer

$$\widehat{\overset{\circ}{j}} = \bigcup \{\beta_D(I) \mid D \in e(j)\}.$$

□

In the next section, we show that $\{\beta_A(I) \mid A \in \mathcal{P}(J)\}$ is a collection of compact semigroups.

2.2. Semigroup Extension. The semigroup $(I, \cup)$ can be extended (uniquely) to the right topological semigroup $(\beta I, \uplus)$ [see [**3**], Chapter 4 and the discussion after Definition 5.24 on page 98]. The terminology *right topological semigroup* means that the binary operation $\uplus$ on βI is associative and for the S-C topology τ on βI [see Remark 2], each function in the collection

$$\{\rho_q \mid q \in \beta I\}$$

is τ-continuous, where $\rho_q(p) = p \uplus q$ for $p, q \in \beta I$. The binary operation $\uplus$ can be characterize as follows. For $B \subseteq I$ and $p, q \in \beta I$, $B \in p \uplus q$ if and only if

$$\{i \in I \mid \{k \in I \mid i \cup k \in B\} \in q\} \in p \tag{2.1}$$

[see [**3**], Theorem 4.12 (b) on page 76]. We need to define notation so that we can handle expression (2.1) more efficiently. For $B \subseteq I$ and $i \in I$, let

$$\lambda_i^{-1}[B] = \{k \in I \mid \lambda_i(k) \in B\},$$

where $\lambda_i(k) = i \cup k$ for $k \in I$, i.e.,

$$\lambda_i^{-1}[B] = \{k \in I \mid i \cup k \in B\}.$$

DEFINITION 7 (Diamond Sets). *For* $q \in \beta I$ *and* $B \subseteq I$, *define*

$$B^{\blacklozenge}(q) = \{\ i \in I \ \mid\ \lambda_i^{-1}[B] \in q\ \}.$$

Using the above definition, we can restate the characterization of $\uplus$ as follows. For $p, q \in \beta I$ and $B \subseteq I$,

$$B \in p \uplus q \quad \text{if and only if} \quad B^{\blacklozenge}(q) \in p. \tag{2.2}$$

REMARK 3. *For* $q \in \beta I$ *and* $B \subseteq I$, *we have*

$$[I \backslash B]^{\blacklozenge}(q) = I \backslash \left[B^{\blacklozenge}(q)\right].$$

Indeed, $i \in [I\backslash B]^{\blacklozenge}(q)$ if and only if $\lambda_i^{-1}[I\backslash B] \in q$ if and only if $I\backslash\lambda_i^{-1}[B] \in q$ (since $\lambda_i^{-1}[I\backslash B] = I\backslash\lambda_i^{-1}[B]$) if and only if $\lambda_i^{-1}[B] \notin q$ (since q is an ultrafilter) if and only if $i \notin B^{\blacklozenge}(q)$ if and only if $i \in I\backslash\left[B^{\blacklozenge}(q)\right]$.

We will now examine examples of diamond sets that will be used in key arguments of this paper. Specifically, for $q \in \beta I$ and $j \in J$ we will consider $\mathring{j}^{\blacklozenge}(q)$ and $\left[I\backslash\mathring{j}\right]^{\blacklozenge}(q)$. Note that $q \in \beta_A(I)$ for some (unique) $A \subseteq J$ by Proposition 1. Since $i \in \mathring{j}$ implies $i \cup k \in \mathring{j}$ for any $k \in I$ and $i \notin \mathring{j}$ implies $i \cup k \in \mathring{j}$ only for $k \in \mathring{j}$, we have

$$\lambda_i^{-1}\left[\mathring{j}\right] = \begin{cases} I, & \text{for } i \in \mathring{j} \\ \mathring{j}, & \text{for } i \notin \mathring{j} \end{cases}. \tag{2.3}$$

PROPOSITION 3. *Let* $B \subseteq J$ *and* $q \in \beta_B(I)$.

(1) *If* $j \in B$, *then* $\mathring{j}^{\blacklozenge}(q) = I$.
(2) *If* $j \in J\backslash B$, *then* $\mathring{j}^{\blacklozenge}(q) = \mathring{j}$.
(3) *If* $j \in B$, *then* $\left[I\backslash\mathring{j}\right]^{\blacklozenge}(q) = \varnothing$.
(4) *If* $j \in J\backslash B$, *then* $\left[I\backslash\mathring{j}\right]^{\blacklozenge}(q) = I\backslash\mathring{j}$.

PROOF. (1) Assume that $j \in B$. Since $q \in \beta_B(I)$ implies $I, \mathring{j} \in q$, we infer $\lambda_i^{-1}\left[\mathring{j}\right] \in q$ for each $i \in I$ from expression (2.3). Hence, $\mathring{j}^{\blacklozenge}(q) = I$.

(2) Assume that $j \in J\backslash B$. Since $\lambda_i^{-1}\left[\mathring{j}\right] = I \in q$ only for $i \in \mathring{j}$ (by expression (2.3) and the fact that $\mathring{j} \notin q$), we infer $\mathring{j}^{\blacklozenge}(q) = \mathring{j}$.

Statements (3) and (4) follows from Remark 3 and statements (1) and (2) respectively. $\square$

PROPOSITION 4. *Let* $A, B \in \mathcal{P}(J)$. *If* $p \in \beta_A(I)$ *and* $q \in \beta_B(I)$, *then* $p \uplus q \in \beta_{B\cup A}(I)$. *In particular, if* $B \subseteq A$, *then* $p \uplus q \in \beta_A(I)$ *and* $q \uplus p \in \beta_A(I)$.

PROOF. Let $j \in B \cup A$. If $j \in B$, then $\mathring{j}^{\blacklozenge}(q) = I$ by Proposition 3 (1). If $j \in A \backslash B$, then $\mathring{j}^{\blacklozenge}(q) = \mathring{j}$ by Proposition 3 (2). In either case, $\mathring{j}^{\blacklozenge}(q) \in p$; therefore, $\mathring{j} \in p \uplus q$ by expression (2.2).

Now let $j \in J \backslash (B \cup A)$. Hence, $\left[I \backslash \mathring{j}\right]^{\blacklozenge}(q) = I \backslash \mathring{j}$ by Proposition 3 (4) (since $J \backslash (B \cup A) \subseteq J \backslash B$). Also, $J \backslash (B \cup A) \subseteq J \backslash A$ implies $I \backslash \mathring{j} \in p$ (since $p \in \beta_A(I)$), which implies $\left[I \backslash \mathring{j}\right]^{\blacklozenge}(q) \in p$, which implies $I \backslash \mathring{j} \in p \uplus q$ by expression (2.2).

We infer $\mathcal{G}_{B \cup A} \subseteq p \uplus q$, which implies $p \uplus q \in \beta_{B \cup A}(I)$. □

PROPOSITION 5. *For $A \subseteq J$, $(\beta_A(I), \uplus)$ is a compact subsemigroup of $(\beta I, \uplus)$.*

PROOF. Note that $p, q \in \beta_A(I)$ implies $p \uplus q \in \beta_{A \cup A}(I) = \beta_A(I)$ by Proposition 4. Consequently, $(\beta_A(I), \uplus)$ is a semigroup.

Observe,

$$\beta_A(I) = \bigcap \left\{ \widehat{E} \mid E \in \mathcal{G}_A \right\},$$

which implies $\beta_A(I)$ is compact, since $\widehat{E}$ is compact for each $E \in \mathcal{G}_A$. □

THEOREM 1. *Let $A \subseteq J$ and let*

$$\mathcal{V}_A = \bigcup \{\beta_B(I) \mid B \subseteq A\}.$$

(a): *$\mathcal{V}_A$ is a closed subsemigroup of βI, $\beta_A(I)$ is an ideal of $\mathcal{V}_A$ and $\beta_A(I) \neq \mathcal{V}_A$ for $A \neq \varnothing$.*

(b): *If $\Gamma \subseteq \beta I$ is a subsemigroup such that $\beta_A(I)$ is an ideal of Γ, then $\Gamma \subseteq \mathcal{V}_A$.*

PROOF. (a) If $p, q \in \mathcal{V}_A$, then $p \in \beta_{B_1}(I)$ and $q \in \beta_{B_2}(I)$ for some $B_1, B_2 \in \mathcal{P}(A)$. Hence, $p \uplus q, q \uplus p \in \beta_{B_1 \cup B_2}(I)$ by Proposition 4, which implies $p \uplus q, q \uplus p \in \mathcal{V}_A$. We infer $\mathcal{V}_A$ is a subsemigroup of βI. Let $p \in \beta_A(I)$ and let $q \in \mathcal{V}_A$. Hence, $q \in \beta_B(I)$ for some $B \subseteq A$. Therefore, $p \uplus q, q \uplus p \in \beta_A(I)$ by Proposition 4 (since $B \cup A = A \cup B = A$). We infer $\beta_A(I)$ is an ideal of $\mathcal{V}_A$. Also, $\beta_A(I) \neq \mathcal{V}_A$ for $A \neq \varnothing$ by Proposition 1.

Let $p \in \beta I \backslash \mathcal{V}_A$. Hence, there exists $D \subseteq J$ such that $p \in \beta_D(I)$ by Proposition 1. Let $j \in D \backslash A$. Hence, $\mathring{j} \in p$, which implies $p \in \widehat{\mathring{j}}$. Let $q \in \widehat{\mathring{j}}$; consequently, $\mathring{j} \in q$. If $q \in \mathcal{V}_A$, then we would have $q \in \beta_B(I)$ for some $B \subseteq A$, which would imply $I \backslash \mathring{j} \in q$ [since $j \notin B$], which would contradict $\mathring{j} \in q$. Therefore, $\widehat{\mathring{j}} \subseteq \beta I \backslash \mathcal{V}_A$. We infer that $\mathcal{V}_A$ is closed.

(b) Let $q \in \Gamma$. Hence, $q \in \beta_D(I)$ for some $D \subseteq J$. So, for $p \in \beta_A(I)$, we have $p \uplus q \in \beta_A(I)$ [since $\beta_A(I)$ is an ideal of Γ] and $p \uplus q \in \beta_{D \cup A}(I)$ [by Proposition 4]. Therefore, $A = D \cup A$ by Proposition 1, which implies $\beta_D(I) \subseteq \mathcal{V}_A$. We infer $\Gamma \subseteq \mathcal{V}_A$. □

COROLLARY 1. *$\beta_J(I)$ is a* ***compact*** *ideal of $(\beta I, \uplus)$.*

PROOF. Note that $\mathcal{V}_J = \beta I$ by Proposition 1. Therefore, apply Proposition 5 and Theorem 1 (a), with $A = J$. □

3. Smallest Ideal of βI

For $A \subseteq J$, Theorem 1 established that $\beta_A(I)$ is a compact ideal of $\mathcal{V}_A$. In this section, we present a proof of Sabine Koppelberg's result that $\beta_J(I)$ is the smallest ideal of βI (see [**4**], Theorem 2.4). First, we need to establish some technical lemmas.

For $i \in I$, let

$$U_i = \bigcap_{j \in i} \mathring{j}. \tag{3.1}$$

LEMMA 1. *Let* $i \in I$. *If* $B \subseteq U_i$, *then* $\lambda_i[B] = B$.

PROOF. If $k \in \lambda_i[B]$, then $k = \lambda_i(b) = i \cup b$ for some $b \in B$. Note that $B \subseteq U_i$ implies $b \in U_i$, which implies $i \subseteq b$, which implies $k = i \cup b = b$, which implies $k \in B$. Therefore, $\lambda_i[B] \subseteq B$. Conversely, if $k \in B$, then $k \in U_i$, which implies $i \subseteq k$, which implies $k = i \cup k = \lambda_i(k)$, which implies $k \in \lambda_i[B]$. Therefore, $B \subseteq \lambda_i[B]$. We infer $\lambda_i[B] = B$. □

Since $\uplus$ is the unique extension of $\cup$ to βI, it can be shown that for $i \in I$ and $p \in \beta I$, we have

$$e(i) \uplus p = \rho_p(e(i)) = \lambda_i[p] \tag{3.2}$$

[see [**3**], Lemma 3.30 on page 58].

LEMMA 2. *Let* $A \subseteq J$ *and let* $p \in \beta_A(I)$. *If* $i \in I$ *such that* $i \subseteq A$, *then* $\lambda_i[p] = p$.

PROOF. Since p, $\lambda_i[p] \in \beta I$ [i.e., both p and $\lambda_i[p]$ (the image filter of p under λ_i) are ultrafilters], it suffices to show that $\lambda_i[p] \subseteq p$. Note that $\{\lambda_i[F] \mid F \in p\}$ is a filter basis for $\lambda_i[p]$. Therefore, it suffices to show that $\lambda_i[F] \in p$ for each $F \in p$. So, let $F \in p$. Note that $U_i \in p$ [by expression (3.1), since $p \in \beta_A(I)$ and $i \subseteq A$] and $\lambda_i[U_i] = U_i$ [Lemma 1] imply $U_i \in \lambda_i[p]$. Also, $B = U_i \cap F \in p$ and $B \subseteq F$, which imply $\lambda_i[B] \subseteq \lambda_i[F]$, which implies $B \subseteq \lambda_i[F]$ [since $\lambda_i[B] = B$ by Lemma 1], which implies $\lambda_i[F] \in p$ [since $B \in p$]. □

PROPOSITION 6. (*Koppelberg*) *Let* $A \subseteq J$. *Let* $A \subseteq J$. *If* $p \in \beta_A(I)$, *then* $p = q \uplus p$ *for each* $q \in \widehat{\mathcal{P}_f(A)}$.

PROOF. Let $A \subseteq J$ and let $p \in \beta_A(I)$. Let $B = \widehat{\mathcal{P}_f(A)}$ and let $\zeta_p : \beta I \to \beta I$ such that $\zeta_p(q) = p$ for each $q \in \beta I$. Note that ζ_p is continuous [with respect to the *Stone-Čech* topology on βI]. If $i \in \mathcal{P}_f(A)$, then $i \in I$ such that $i \subseteq A$, which implies $\lambda_i[p] = p$ [Lemma 2], which implies

$$\rho_p(e(i)) = \lambda_i[p] = p = \zeta_p(e(i))$$

[see expression (3.2)]. Hence, the continuous functions ρ_p and ζ_p agree on $e[\mathcal{P}_f(A)]$; therefore, ρ_p and ζ_p agree on $\widehat{\mathcal{P}_f(A)}$ [since $\widehat{\mathcal{P}_f(A)}$ is the closure of $e[\mathcal{P}_f(A)]$ (see Remark 1)]. Therefore, for $q \in \widehat{\mathcal{P}_f(A)}$, we have

$$p = \zeta_p(q) = \rho_p(q) = q \uplus p.$$

□

COROLLARY 2. *If* $p \in \beta_J(I)$, *then* $p = q \uplus p$ *for each* $q \in \beta I$.

PROOF. If $p \in \beta_J(I)$, then $p = q \uplus p$ for each $q \in \beta I = \widehat{\mathcal{P}_f(J)}$ by Proposition 6. $\square$

THEOREM 2. (*Koppelberg*) *Let J be an infinite set and $I = \mathcal{P}_f(J)$. For semigroup $(\beta I, \uplus)$, $K(\beta I) = \beta_J(I)$ [i.e., $\beta_J(I)$ is the smallest ideal of $(\beta I, \uplus)$].*

PROOF. Since $\beta_J(I)$ is an ideal of βI [Corollary 1], it suffices to show that $\beta_J(I) \subseteq M$ for any ideal M of βI. So, let M be an ideal of βI. Let $p \in \beta_J(I)$. Since $M \neq \varnothing$ [by definition], there exists $q \in \beta I$ such that $q \in M$. Note that $q \uplus p \in M$ [since M is a right ideal of βI]. Therefore, $p \in M$, since $p = q \uplus p$ by Corollary 2. Consequently, $\beta_J(I) \subseteq M$. $\square$

REMARK 4. *For $A \subseteq J$, it can be shown that $\beta\mathcal{P}_f(A)$ [the collection of all ultrafilters on $\mathcal{P}_f(A)$] is canonically homeomorphic and isomorphic to $\widehat{\mathcal{P}_f(A)}$ and that $\beta_A(\mathcal{P}_f(A))$ is canonically homeomorphic and isomorphic to $\beta_A(I) \cap \widehat{\mathcal{P}_f(A)}$.*

Since $K(\beta\mathcal{P}_f(A)) = \beta_A(\mathcal{P}_f(A))$ [Theorem 2], we have:

COROLLARY 3. (*Koppelberg*) *For $A \subseteq J$, if $A \neq \varnothing$, then $K\left(\widehat{\mathcal{P}_f(A)}\right) = \beta_A(I) \cap \widehat{\mathcal{P}_f(A)}$ [i.e., $\beta_A(I) \cap \widehat{\mathcal{P}_f(A)}$ is the smallest ideal of the semigroup $\left(\widehat{\mathcal{P}_f(A)}, \uplus\right)$].*

4. Ideals of βI

Consider $A \subseteq J$. As noted in the Preliminaries, $\mathcal{V}_A$ is a compact, subsemigroup of $(\beta I, \uplus)$. Also, *in general*, $\mathcal{V}_A$ is not an ideal of $(\beta I, \uplus)$; however, if $A \neq J$, then from Proposition 1, we infer that $\beta I \backslash \mathcal{V}_A$ is an open, ideal of $(\beta I, \uplus)$. Again, using Proposition 1, we make the following observations: (1) $\beta I \backslash \mathcal{V}_A = \bigcup \{\beta_D(I) \mid D \in \mathcal{P}(J) \backslash \mathcal{P}(A)\}$ and, if A is a non empty, proper subset of J, then $\mathcal{V}_{J\backslash A} \backslash \beta_\varnothing(I)$ is a proper subset of $\beta I \backslash \mathcal{V}_A$; (2) if $E \subseteq J$, then $\mathcal{V}_E \backslash \mathcal{V}_A = \bigcup \{\beta_D(I) \mid D \in \mathcal{P}(E) \backslash \mathcal{P}(A)\}$ and

$$\mathcal{V}_E \cap \mathcal{V}_A = \mathcal{V}_{E \cap A},$$

which implies $\mathcal{V}_A \cap \mathcal{V}_{J\backslash A} = \beta_\varnothing(I)$. Also,

$$\mathcal{V}_E \cup \mathcal{V}_A = \bigcup \{\beta_D(I) \mid D \in \mathcal{P}(E) \cup \mathcal{P}(A)\}.$$

Note that if $A \backslash E \neq \varnothing$ and $E \backslash A \neq \varnothing$, then $\mathcal{V}_E \cup \mathcal{V}_A$ is a proper subset of $\mathcal{V}_{E \cup A}$. Indeed, if $a \in A \backslash E$ and $e \in E \backslash A$, then $\beta_{\{a,e\}}(I) \subseteq \mathcal{V}_{E \cup A}$ and Proposition 1 implies that $\beta_{\{a,e\}}(I) \cap \mathcal{V}_A = \varnothing = \beta_{\{a,e\}}(I) \cap \mathcal{V}_E$. Observe that Propositions 1 and 4 imply that, for $A \subseteq J$ and $E \subseteq J$, $\mathcal{V}_E \backslash \mathcal{V}_A$ is an ideal of $(\mathcal{V}_E, \uplus)$ that is open with respect to the topology induced on $\mathcal{V}_E$ by the the Stone-Čech Compactification topology on βI.

Observe that if $A \subseteq J$ and $J \backslash A$ is finite, then $B \subseteq A$ implies $\{I \backslash \mathcal{P}_f(A)\} \cup \mathcal{G}_B$ does not have the finite intersection property. Indeed, $i \in I \backslash \mathcal{P}_f(A)$ implies $b \in i$ for some $b \in J \backslash A$, which implies $i \notin \bigcap_{j \in J \backslash A} I \backslash \mathring{j}$. So, if $J \backslash A$ is finite, then $I \backslash \mathcal{P}_f(A) \notin q$ for any $q \in \mathcal{V}_A$, which implies $\mathcal{P}_f(A) \in q$ for any $q \in \mathcal{V}_A$, which implies $\mathcal{V}_A \subseteq \widehat{\mathcal{P}_f(A)}$. Also, if $B \subseteq J$ and $A \subseteq J$ such that $\{\mathcal{P}_f(A)\} \cup \mathcal{G}_B$

has the finite intersection property, then $B \subseteq A$. So, from Proposition 1, we infer $\widehat{\mathcal{P}_f(A)} \subseteq \mathcal{V}_A$ for any $A \subseteq J$. Therefore, if $A \subseteq J$ and $J \backslash A$ is finite, then

$$\mathcal{V}_A = \widehat{\mathcal{P}_f(A)}. \tag{4.1}$$

In fact, it can be shown that $\mathcal{V}_A$ is clopen if and only if $J \backslash A$ is finite [see [**1**], Corollary 3.6 and Remark 3.7], since $\mathcal{V}_A = \widehat{E}$ for some $E \subseteq I$ implies $E = \mathcal{P}_f(A)$. Consequently, for $B \subseteq A$, $J \backslash A$ is infinite if and only if $\beta_B(I) \cap [I \backslash \widehat{\mathcal{P}_f(A)}] \neq \varnothing$ [again, see [**1**], Corollary 3.6 and Remark 3.7].

PROPOSITION 7. *Let* $A \subseteq J$. *If* $J \backslash A$ *is finite, then* $K(\mathcal{V}_A) = \beta_A(I)$ [*i.e.,* $\beta_A(I)$ *is the smallest ideal of* $\mathcal{V}_A$].

PROOF. Using expression (4.1) and Koppelberg's result [i.e., Corollary 3], we have

$$K(\mathcal{V}_A) = K\left(\widehat{\mathcal{P}_f(A)}\right) = \beta_A(I) \cap \widehat{\mathcal{P}_f(A)} = \beta_A(I) \cap \mathcal{V}_A = \beta_A(I).$$

□

To see that the converse of Proposition 7 is true requires a bit more machinery.

DEFINITION 8. *Let* J *be an infinite set and* $I = \mathcal{P}_f(J)$. *For* $C \subseteq J$, *let* $I_C = \{i \in I \mid i \cap C \neq \varnothing\}$. *Define the collection*

$$\mathcal{W}_J = \{I_C \mid C \subseteq J \quad \text{is infinite}\}.$$

Observe that each element of $\mathcal{W}_J$ is an **ideal** of $(I, \cup)$ and $I_C = I \backslash \mathcal{P}_f(J \backslash C)$.

REMARK 5. *It can be shown that if* $A \subseteq J$, *then* $\mathcal{W}_J \cup \mathcal{G}_A \cup \mathcal{H}$ *has the finite intersection property.*

[see [**1**], Proposition 3.5]

DEFINITION 9. *Let* J *be an infinite set and* $I = \mathcal{P}_f(J)$. *For* $A \subseteq J$, *define*

$$\mathcal{X}_A = \mathcal{W}_J \cup \{V_n \mid n \in \mathbb{N}\} \cup \{U_i \mid i \in \mathcal{P}_f(A)\},$$

where $U_i = \bigcap_{j \in i} \mathring{j}$ *for* $i \in I$ *and* $V_n = \bigcap_{t=1}^{n} I \backslash [J]^t$ *for* $n \in \mathbb{N}$.

Note that, for $A \subseteq J$, $\mathcal{X}_A$ is a collection of **ideals** of $(I, \cup)$ and for $D \subseteq J$, if $A \subseteq D$, then $\mathcal{X}_A \subseteq \mathcal{X}_D$.

REMARK 6. *It can be shown that* (*1*) *if* A *and* D *are subsets of* J *such that* $A \subseteq D$, *then* $\mathcal{X}_A \cup \mathcal{G}_D$ *has the finite intersection property and* (*2*) *if* B *is a proper subset of* A, *then* $\mathcal{X}_A \cup \mathcal{G}_B$ *does not have the finite intersection property.*

[see [**1**], Corollary 3.10 and Remark 3.11]

DEFINITION 10. *Let* J *be an infinite set and let* $I = \mathcal{P}_f(J)$. *For* $A \subseteq J$, *define*

$$\Upsilon_A = \bigcap \left\{ \widehat{E} \mid E \in \mathcal{X}_A \right\}$$

and

$$\Theta_A = \bigcap \left\{ \widehat{E} \mid E \in \mathcal{X}_A \cup \mathcal{G}_A \right\}.$$

For $A \subseteq J$, the sets Υ_A and Θ_A are closed subsets of βI, since both Υ_A and Θ_A are the intersections of collections of closed subsets of βI. Observe that $\Theta_A \subseteq \Upsilon_A$ and Remark 6 (1) implies $\Theta_A \neq \varnothing$. Since $\mathcal{X}_A$ is a collection of ideals of $(I, \cup)$, we have that Υ_A is a **closed ideal** of βI. Also, since $\mathcal{X}_J \cup \mathcal{G}_J = \mathcal{X}_J$, we have

$$\Theta_J = \Upsilon_J = \beta_J(I). \tag{4.2}$$

In addition, if $D \subseteq J$ such that $A \subseteq D$, then $\Upsilon_D \subseteq \Upsilon_A$. Recall that $A \subseteq J$ implies

$$\beta_A(I) = \bigcap\left\{\widehat{E} \mid E \in \mathcal{G}_A\right\} \quad \text{and} \tag{4.3}$$
$$\beta_0^{(A)}(I) = \bigcap\left\{\widehat{E} \mid E \in \mathcal{G}_A \cup \mathcal{H}\right\}.$$

From Definitions 9, 10, Remark 6 and expression (4.3), we infer

$$\Theta_A = \beta_A(I) \cap \Upsilon_A = \beta_0^{(A)}(I) \cap \Upsilon_A, \tag{4.4}$$

which implies

$$\Theta_A \text{ is a } \textbf{closed ideal} \text{ of } \beta_A(I) \text{ and } \beta_0^{(A)}(I) \tag{4.5}$$

[see [**1**], Proposition 4.2].

Recall that if A is *infinite*, then $\beta_A(I) = \beta_0^{(A)}(I)$ [see [**2**], Proposition 4.2].

REMARK 7. *The collection* $\{\Theta_A \mid A \subseteq J\}$ *is pairwise disjoint.*

Indeed, for subsets A and B of J such that $A \neq B$, we have

$$\Theta_A \cap \Theta_B \subseteq \beta_A(I) \cap \beta_B(I) = \varnothing$$

by Proposition 1 and expression 4.4.

REMARK 8. *Let A and D be subsets of J. If A is a proper subset of D, then*

$$\mathcal{V}_A \cap \Theta_D = \varnothing.$$

Indeed, $B \subseteq A$ implies $B \neq D$, which implies $\beta_B(I) \cap \Theta_D \subseteq \beta_B(I) \cap \beta_D(I) = \varnothing$ by Proposition 1. Therefore,

$$\mathcal{V}_A \cap \Theta_D = \left[\bigcup_{B \subseteq A} \beta_B(I)\right] \cap \Theta_D = \bigcup_{B \subseteq A} \left[\beta_B(I) \cap \Theta_D\right] = \varnothing.$$

PROPOSITION 8. *If $A \subseteq J$, then $\Upsilon_A = \bigcup\{\,\Theta_D \mid A \subseteq D\,\}$.*

PROOF. Note that for $A \subseteq D$, we have $\Theta_D \subseteq \Upsilon_D \subseteq \Upsilon_A$ [see expression (4.4)]. Therefore,

$$\bigcup\{\Theta_D \mid A \subseteq D\} \subseteq \Upsilon_A. \tag{4.6}$$

Conversely, assume that $p \in \Upsilon_A$. Hence, $p \in \beta_{D_0}(I)$ for some (unique) $D_0 \subseteq J$ [Proposition 1], which implies $\mathcal{G}_{D_0} \subseteq p$ [by definition of $\beta_{D_0}(I)$], which implies $\{U_i \mid i \in \mathcal{P}_f(D_0)\} \subseteq p$. Note that $p \in \Upsilon_A$ implies $\mathcal{W}_J \cup \{V_n \mid n \in \mathbb{N}\} \subseteq \mathcal{X}_A \subseteq p$. So, $\mathcal{W}_J \cup \{V_n \mid n \in \mathbb{N}\} \subseteq p$ and $\{U_i \mid i \in \mathcal{P}_f(D_0)\} \subseteq p$ imply $\mathcal{X}_{D_0} = \mathcal{W}_J \cup \{V_n \mid n \in \mathbb{N}\} \cup \{U_i \mid i \in \mathcal{P}_f(D_0)\} \subseteq p$. Also, $\mathcal{X}_{D_0} \subseteq p$ and $\mathcal{G}_{D_0} \subseteq p$

imply $\mathcal{X}_{D_0} \cup \mathcal{G}_{D_0} \subseteq p$, which implies $p \in \widehat{E}$ for each $E \in \mathcal{X}_{D_0} \cup \mathcal{G}_{D_0}$, which implies

$$p \in \bigcap \left\{ \widehat{E} \mid E \in \mathcal{X}_{D_0} \cup \mathcal{G}_{D_0} \right\} = \Theta_{D_0}. \tag{4.7}$$

Next, we will show $A \subseteq D_0$. So, let $j \in A$. Hence, $\{j\} \in \mathcal{P}_f(A)$, which implies $\mathring{j} = U_{\{j\}} \in \mathcal{X}_A$ [since $\{U_i \mid i \in \mathcal{P}_f(A)\} \subseteq \mathcal{X}_A$]. Note that $p \in \Upsilon_A$ implies $\mathcal{X}_A \subseteq p$. Hence, $\mathring{j} \in p$.

If $j \notin D_0$, then we would have $j \in J \backslash D_0$, which would imply $I \backslash \mathring{j} \in p$ [since $p \in \beta_{D_0}(I)$], which would contradict $\mathring{j} \in p$. We infer $A \subseteq D_0$. Hence,

$$\Theta_{D_0} \subseteq \bigcup \{\Theta_D \mid A \subseteq D\}, \tag{4.8}$$

which implies

$$p \in \bigcup \{\Theta_D \mid A \subseteq D\} \tag{4.9}$$

[by expressions (4.7) and (4.8)].

We infer

$$\Upsilon_A = \bigcup \{\Theta_D \mid A \subseteq D\}.$$

□

COROLLARY 4. *If* $A \subseteq J$, *then* $\Upsilon_A \cap \mathcal{V}_A = \Theta_A$.

PROOF. Observe,

$$\Upsilon_A \cap \mathcal{V}_A = \bigcup_{A \subseteq D} [\Theta_D \cap \mathcal{V}_A] = \Theta_A \cap \mathcal{V}_A = \Theta_A$$

by Proposition 8 and Remark 8, since $\Theta_A \subseteq \beta_A(I) \subseteq \mathcal{V}_A$ [see expression (4.4)]. □

COROLLARY 5. *If* $A \subseteq J$, *then* Θ_A *is a closed ideal of* $\mathcal{V}_A$.

PROOF. Corollary 4, Υ_A is a closed ideal of βI and $\mathcal{V}_A$ is a closed subsemigroup of βI. □

Recall that A being a proper subset of J implies $I \backslash \mathcal{P}_f(A)$ is an ideal of $(I, \cup)$, which implies $\widehat{I \backslash \mathcal{P}_f(A)}$ is a closed ideal of $(\beta I, \uplus)$. Note that if $A \subseteq J$ such that $J \backslash A$ is infinite, then, for $B \subseteq A$, we have

$$\Theta_B \subseteq \beta_0^{(B)}(I) \bigcap [\widehat{I \backslash \mathcal{P}_f(A)}]. \tag{4.10}$$

Indeed, $\Theta_B \subseteq \beta_0^{(B)}(I)$ by expression (4.5) and note that $I \backslash \mathcal{P}_f(A) \in \mathcal{W}_J \subseteq \mathcal{X}_B \cup \mathcal{G}_B$ [since $I_{J \backslash A} = I \backslash \mathcal{P}_f(A)$ (see Definition 8)]. Therefore,

$$\Theta_B = \bigcap \left\{ \widehat{E} \mid E \in \mathcal{X}_B \cup \mathcal{G}_B \right\} \subseteq [\widehat{I \backslash \mathcal{P}_f(A)}].$$

PROPOSITION 9. *Let* $A \subseteq J$ *and* $B \subseteq A$. *If* A *and* $J \backslash A$ *are infinite, then* Θ_B *is a proper closed ideal of* $\beta_0^{(B)}(I)$.

PROOF. Note that A being infinite implies $\beta_0^{(B)}(I) \bigcap \widehat{\mathcal{P}_f(A)} \neq \varnothing$ [see Definition 6 and [**1**], Proposition 3.1]. Also, $J \backslash A$ being infinite implies $\Theta_B \subseteq \beta_0^{(B)}(I) \bigcap [\widehat{I \backslash \mathcal{P}_f(A)}]$ by expression (4.10). Consequently, Θ_B is a proper closed ideal of $\beta_0^{(B)}(I)$ [see expression (4.5)]. □

Let $B \subseteq J$. Since $\beta_0^{(B)}(I)$, $\beta_B(I)$ and $\mathcal{V}_B$ are compact subsemigroups of $(\beta I, \uplus)$, we have that $K(\mathcal{V}_B)$, $K(\beta_B(I))$ and $K\left(\beta_0^{(B)}(I)\right)$ exist [see [**3**], Corollary 2.6 and Theorem 1.51]. The next proposition implies the converse of Proposition 7.

PROPOSITION 10. *Let* $A \subseteq J$ *and* $B \subseteq A$. *If* A *and* $J \backslash A$ *are infinite, then* $\overline{K(\mathcal{V}_B)}$ *is a proper subset of* $\beta_0^{(B)}(I)$ *and*

$$K(\mathcal{V}_B) = K(\beta_B(I)) = K\left(\beta_0^{(B)}(I)\right).$$

PROOF. Note that Corollary 5 implies $\overline{K(\mathcal{V}_B)} \subseteq \Theta_B$, which implies $\overline{K(\mathcal{V}_B)}$ is a proper subset of $\beta_0^{(B)}(I)$ by Proposition 9. Recall that $\beta_0^{(B)}(I)$ and $\beta_B(I)$ are compact subsemigroups of the compact semigroup $(\mathcal{V}_B, \uplus)$ and $\beta_0^{(B)}(I) \subseteq \beta_B(I)$. Therefore,

$$K(\beta_B(I)) = K(\mathcal{V}_B) \cap \beta_B(I) = K(\mathcal{V}_B)$$

and

$$K\left(\beta_0^{(B)}(I)\right) = K(\mathcal{V}_B) \cap \beta_0^{(B)}(I) = K(\mathcal{V}_B)$$

[see [**3**], Theorem 1.65 and Corollary 2.6]. □

References

[1] A. D. Grainger, *Ideals of Ultrafilters on the Collection of Finite Subsets of an Infinite Set,* Semigroup Forum, Vol. 73 (2006), pp. 234-242.

[2] A. D. Grainger, *Ultrafilters on the Collection of Finite Subsets of an Infinite Set,* Semigroup Forum, Vol. 67 (2003), pp. 443-453.

[3] Neil Hindman and Dona Strauss, *Algebra in the Stone-Čech Compactification*, Walter de Gruyter & Co., New York 1998.

[4] Sabine Koppelberg, *The Stone-Čech Compactification of a Semilattice*, Semigroup Forum, Vol. 72, No. 1 (2006), pp. 63-74.

DEPARTMENT OF MATHEMATICS, MORGAN STATE UNIVERSITY, BALTIMORE, MARYLAND 21251
E-mail address: agrainge@jewel.morgan.edu

Contemporary Mathematics
Volume **467**, 2008

The Atlas of Lie Groups and Representations: Scope and Successes

Alfred G. Noël

To Fokko du Cloux in memoriam

Abstract. The Atlas of Lie Groups and Representations project's main goal is to compute the Unitary Dual of a given reductive real Lie group from the input of the root datum corresponding to its complexification. An earlier version of the software was written by the late Fokko du Cloux. That version computes the representation theories of all the real forms of a given complex reductive Lie group and provides an implementation of the Kazhdan-Lusztig-Vogan algorithm that was used to compute the KLV polynomials for the split form of E_8, a result whose announcement in March 2007 was intensely covered by both the national and international media.

My own work within this international group involves developing and implementing algorithms for computing the Unitary Dual and determining special nilpotent orbits attached to Coxeter cells. I work very closely with David Vogan at MIT and Steven Glenn Jackson at the University of Massachusetts Boston. I will present some of the main ideas behind the current version of the software. I will provide some examples also. This paper is in memory of Fokko du Cloux, the mathematician who was the lead developer. He died in November 2006.

1. Introduction

In 2002, Jeff Adams of the University of Maryland was thinking about the possibility of making calculations for real Lie groups that would be similar to those that can be carried out in LiE [**VL**] for complex Lie groups, his ultimate goal being the computation of the *Unitary Dual* of a given real reductive Lie group. This is very ambitious. He contacted several mathematicians including David Vogan, Peter Trapa, Bill Casselman, Dan Barbasch and Jonn Stembridge with the idea of building a powerful team of theorists and computational experts in order to

Key words and phrases. Lie Groups, Kazhdan-Lusztig-Vogan polynomials, Unitary Dual.

The author was partially supported by NSF grant #DMS 0554278 .

Most of this work was done while the author was a visiting scholar in the Mathematics department at the Massachusetts Institute of Technology from February 2006 to August 2007. He thanks his host David Vogan for many stimulating discussions . He also thanks all the other members of the Atlas of Lie groups and representations team for having taught him so much in such a short time.

tackle this monumental task. Very soon he was urged to contact Fokko du Cloux, a world renowned expert in Coxeter groups computation, who at that time, had already built the best program for computing Kazhdan-Lusztig polynomials for such groups [**F**]. Fokko began to work on what is now known as the *Atlas of Lie groups and Representations* package around March 2003. By November 2005, at the time he was diagnosed with ALS, he managed to implement all the algorithms that were needed to compute the character of a reductive Lie group. This is even more remarkable given the fact that Fokko had to learn the extremely difficult mathematics as he progressed and then developed fast and sophisticated algorithms which he implemented himself in C++. Here is a list of the software capabilities:

The user may input:

- Input the data for a general connected complex group G,
- Specify an inner class of strong real forms of G,
- Specify a real form of G,

The software will compute:
- The component group of G,
- The conjugacy classes of Cartan subgroups of G,
- The ("real") Weyl group of reach Cartan subgroup,
- Parameters for the irreducible representations of G with regular integral infinitesimal character,
- Gradings of roots (compact imaginary, non-compact imaginary, parity condition, etc.),
- The cross action and Cayley transforms of irreducible characters,
- Kazhdan-Lusztig polynomials for representations with regular integral infinitesimal character

Information about algorithms, research papers, tutorials and downloading issues is found at http://atlas.math.umd.edu/index.atlas.html.

It would be an impossible task to describe all this capability here. Instead I will first give a summary of the ideas behind the E_8 character table computation. Then I will provide a quick tutorial on the implementation of "Real Weyl groups" that is quite specific to the Atlas. And I will sketch some of the techniques that are being considered in order to complete the computation of the unitary dual. Finally, Jeffrey Adams will say a few words about Fokko and his immense contribution to the Atlas project.

2. Character tables and Kazhdan-Lusztig Polynomials

This section relies heavily on a recent paper of Vogan [**V**] which was the subject of the talk that I delivered at the 13th Conference for African American in the Mathematical Sciences at the University of Massachusetts Boston in June 2007.

DEFINITION 2.1. Let G be a reductive algebraic Lie goup. A *representation* of G on a complex Hilbert space $V \neq 0$ is a homomorphism π of G into the group of bounded linear operators on V such that:

$$G \times V \to V \text{ given by } (g, v) \to \pi(g)v \text{ is continuous}$$

We say that $\dim \pi = \dim V$ and that (π, V) is *unitary* if $\pi(g)$ is a unitary operator. Moreover (π, V) is *irreducible* if the only closed G -invariant subspaces of V are $\{0\}$ and V.

The set of irreducible unitary representations (**The Unitary Dual**) of G denoted by $\hat{G}$ is a fundamental tool to understand the actions of G. For G compact $\hat{G}$ is essentially determined. Barbasch has treated the classical complex groups. However the following cases are still not resolved:

Type A: $SU(p,q)$ for $(p, q > 2)$
Type B: $SO(p,q)$ for $(p, q \geq 3)$
Type C: $Sp(p,q)$ for $(p, q \geq 2)$
Type D: $SO(p,q)$ for $(p, q \geq 3)$, $SO^*(2n)$ for $(n \geq 4)$
Type F_4: $F_4(\mathbb{C})$, $F_4(split)$
Type E_6: $E_6(\mathbb{C})$, $E_6(split)$, $E_6(Hermitian)$, $E_6(quaternionic)$
Type E_7: $E_7(\mathbb{C})$ and all real non-compact forms
Type E_8: $E_8(\mathbb{C})$ and all real non-compact forms

Works of Harish-Chandra and others tell us that in order to describe the full unitary dual we will need to first find the set Π of equivalence classes of irreducible *quasisimple* representations of G. Then describe $\Pi_h \subseteq \Pi$ the equivalence classes of irreducible hermitian representations of G. And finally describe $\hat{G}$ as a subset of Π_h.

The first two steps are addressed in [**K**] and [**L**]. Here is an approximate statement [**V**].

THEOREM 2.2. *(Knapp-Zuckerman, Langlands). There is a natural bijection between Π and a countable discrete collection of complex algebraic variteies $X_i(\mathbb{C})$. Each of these algebraic varities is defined over $\mathbb{R}$, and the subset Π_h corresponds to the real points $X_i(\mathbb{R})$.*

The *character* of a finite dimensional representation π is defined to be:

$$\theta_\pi(g) = tr(\pi(g)).$$

This definition is not valid for infinite dimensional representations and is of no use in the study of non compact real Lie groups for

PROPOSITION 2.3. Let G be a non compact linear semisimple group. Then any nontrivial irreducible unitary representation of G is of infinite dimension.

For infinite dimensional representations the operators $\pi(g)$ are rarely of trace class. Fortunately a theorem of Harish-Chandra [**HC**] gives us a way out of this difficulty by defining θ_π as a *generalized function* (continuous linear functional on spaces of compactly supported smooth measures) on G such that there exists a conjugation-invariant open subset G' of G with the property that $G \setminus G'$ has measure zero, so that θ_π restricted to G' is a conjugation-invariant analytic function θ'_π locally integrable on G.

To write a character table for G one has to write down a formula for each θ'_π. It turns out that θ'_π is a quotient of finite integer combinations of exponential functions. Although there are infinitely many irreducible representations we can, using the Jantzen-Zuckerman *translation principle*, partition them into a finitely many *translation families*.

For a fixed infinitesimal character λ Harish-Chandra computed an explicit basis

$$\{\theta_1^\lambda, \theta_2^\lambda, \dots \theta_N^\lambda\}$$

such that

$$\theta_\pi = \sum_{j=1}^{N} \epsilon(\pi, j) P_{\pi,j}^\lambda(1)\theta_j^\lambda \text{ with } \epsilon(\pi, j) = \pm 1.$$

Here the $P_{\pi,j}^\lambda$'s are Kazhdan-Lusztig-Vogan polynomials. The coefficients of $P_{\pi,j}^\lambda$ are all non negative integers. Moreover the Jantzen-Zuckerman translation principle implies that as λ varies there are only finitely many possible matrices $P_{\pi,j}^\lambda(1)$. Using the Langlands-Zuckerman bijection and an appropriate ordering of the θ_j we can obtain $P_{\pi_i,j}^\lambda(1)$ a lower triangular matrix where π_i is the irreducible representation corresponding to θ_i. Readers who are interested in more details about the computation of the matrix $P_{\pi_i,j}^\lambda(1)$ may want to consult [**V**].

The main point is that for fixed infinitesimal character this is a finite computation. Next we give an example using the Atlas.

```
anoel\% ./atlas.exe
This is the Atlas of Reductive Lie Groups Software Package version
0.2.5 2006/09/26 10:49:58.

Enter "help" if you need assistance.

empty: klbasis
Lie type: C2 sc s
(weak) real forms are:
0: sp(2)
1: sp(1,1)
2: sp(4,R)
enter your choice: 2
possible (weak) dual real forms are:
0: so(5)
1: so(4,1)
2: so(2,3)
enter your choice: 2

Name an output file (hit return for stdout):
Full list of non-zero Kazhdan-Lusztig-Vogan polynomials:

 0:  0: 1

 1:  1: 1

 2:  2: 1
```

```
 3:  3: 1

 4:  0: 1
     2: 1
     4: 1

 5:  1: 1
     3: 1
     5: 1

 6:  0: 1
     1: 1
     6: 1

 7:  0: 1
     1: 1
     2: 1
     3: 1
     4: 1
     5: 1
     6: 1
     7: 1

 8:  0: 1
     1: 1
     2: 1
     4: 1
     6: 1
     8: 1

 9:  0: 1
     1: 1
     3: 1
     5: 1
     6: 1
     9: 1

10:  0: 1
     1: 1
     2: 1
     3: 1
     4: 1
     5: 1
     6: 1
     7: 1
     8: 1
     9: 1
    10: 1
```

```
11:  2: q
     3: q
     7: 1
    11: 1

78 pairs
30 zero polynomials; 48 nonzero polynomials
```

This output displays the rows of the matrix for the infinitesimal charcacter ρ when $G = Sp_4(\mathbb{R})$. The last row for example says that $P(11,2) = P(11,3) = q$, $P(11,7) = P(11,11) = 1$, and $P(11,k) = 0$ for all other k. Furthermore this row gives a formula for the irreducible character of representation number 11:

$$\pi_{11} = \theta_{11} - \theta_7 - \theta_3 - \theta_2.$$

3. Real Weyl Groups

This section is based on a lecture given by Vogan in July 2005 at the American Institute of Mathematics. Let G be the real points of a complex connected reductive algebraic group $G_{\mathbb{C}}$. We describe the real Weyl group of G as implemented in the *Atlas of Lie groups and representations.*

Let $G_{\mathbb{C}}$ be a complex connected reductive algebraic group defined over $\mathbb{R}$ and G the set of real points of $G_{\mathbb{C}}$. Let $H_{\mathbb{C}}$ be a maximal torus of $G_{\mathbb{C}}$ defined over $\mathbb{R}$ and H the set of real points of $H_{\mathbb{C}}$, a Cartan subgroup of G. Denote by $W(G_{\mathbb{C}}, H_{\mathbb{C}})$ the complex Weyl group of $G_{\mathbb{C}}$ defined by $H_{\mathbb{C}}$. Let R be the root system defined by $H_{\mathbb{C}}$. Then $R \subseteq X^*(H_{\mathbb{C}})$, the character lattice of $H_{\mathbb{C}}$. The set of coroots is $R^\vee \subseteq X_*(H_{\mathbb{C}})$, the co-character lattice of $H_{\mathbb{C}}$. If $\alpha \in R$ then s_α, the reflection defined by α acts on $X^*(H_{\mathbb{C}})$ by

$$s_\alpha(\lambda) = \lambda - \langle \lambda, \alpha^\vee \rangle \alpha.$$

It is a fact that $W(G_{\mathbb{C}}, H_{\mathbb{C}})$ is generated by $\{s_\alpha \text{ for } \alpha \text{ simple }\}$. Note also that $W(G_{\mathbb{C}}, H_{\mathbb{C}}) = N_{G_{\mathbb{C}}}(H_{\mathbb{C}})/H_{\mathbb{C}}$ and $W(G,H) = N_G(H)/H$.

Finally, we write $W(G_{\mathbb{C}}, H_{\mathbb{C}})(\mathbb{R})$ to denote the real points of $W(G_{\mathbb{C}}, H_{\mathbb{C}})$. We shall describe these Weyl groups.

Let θ be the Cartan involution of $G_{\mathbb{C}}$ for G preserving $H_{\mathbb{C}}$. Then θ is an algebraic involution of $G_{\mathbb{C}}$ preserving G and $K = G^\theta$, the points of G fixed by θ is a maximal compact subgroup of G. Moreover θ acts on $X^*, R, X_*, R^\vee \ldots.$

A root α is said to be:

real if $\theta(\alpha) = -\alpha$ that is $\alpha_{|_H}$ takes real values.
imaginary if $\theta(\alpha) = \alpha$ that is $\alpha_{|_H}$ takes values in the unit circle.
complex if $\theta(\alpha) \neq \pm\alpha$.

An imaginary root is said to be *compact (non-compact)* if θ acts as 1 (-1) on the corresponding root space in the Lie algebra $\mathfrak{g}_{\mathbb{C}}$ of $G_{\mathbb{C}}$.

Since $H_{\mathbb{C}}$ and $N_{G_{\mathbb{C}}}(H_{\mathbb{C}})$ are both defined over $\mathbb{R}$, the quotient $W(G_{\mathbb{C}}, H_{\mathbb{C}})$ is defined over $\mathbb{R}$ as well. It is not difficult to see that the group of real points is

$W(G_{\mathbb{C}}, H_{\mathbb{C}})(\mathbb{R}) = W(G_{\mathbb{C}}, H_{\mathbb{C}})^{\theta}$, the elements of the complex Weyl group commuting with θ.

3.1. Twisted Involutions. The group $G_{\mathbb{C}}$ can be specified by giving a based root datum $(X^*, R^+, X_*, (R^{\vee})^+)$, where $R^+ \subset X^*$ is a set of positive roots, and $X_* \subset X_*$, the corresponding set of positive coroots.

ATLAS always begins with "an inner class of real forms of $G_{\mathbb{C}}$". This is equivalent to an involutive automorphism τ of the based root datum; τ is an automorphism of $X^* \simeq \mathbb{Z}^n$ of order 1 or 2. This means that τ may be regarded as an $n \times n$ integer matrix. In order to be an automorphism, τ must preserve R^+, and its transpose ${}^t\tau$ must preserves $(R^{\vee})^+$.

A very common and important case is τ equal to the identity.

Internally, ATLAS keeps the semidirect product $W(G_{\mathbb{C}}, H_{\mathbb{C}}) \rtimes \{1, \tau\}$ as the working Weyl group. The automorphism τ permutes the generators of the Coxeter group $W(G_{\mathbb{C}}, H_{\mathbb{C}})$.

A twisted involution θ is an element of order 1 or 2 in the coset $W(G_{\mathbb{C}}, H_{\mathbb{C}}) \cdot \tau$. That is, a twisted involution is an element $w \cdot \tau$ such that $w \in W(G_{\mathbb{C}}, H_{\mathbb{C}})$ and $w\tau(w) = 1$.

Note that an element of order 2 in $W(G_{\mathbb{C}}, H_{\mathbb{C}})$ is a product of reflections in orthogonal (not necessarily simple) roots.

What commutes with θ? First, every reflection in a real or imaginary root.

3.2. Description of Real Weyl Groups. Define:
$R_{real} = R \cap (-1 \text{ eigenlattice of } \theta \text{ on } X^*)$
$R_{im} = R \cap (+1 \text{ eigenlattice of } \theta \text{ on } X^*)$

These are the only reflections in $W(G_{\mathbb{C}}, H_{\mathbb{C}})^{\theta}$. We have

$$W(R_{real}) \times W(R_{im}) \subseteq W(G_{\mathbb{C}}, H_{\mathbb{C}})^{\theta}$$

and by orthogonality $W(R_{real})$ and $W(R_{im})$ commute.

Let $2\rho_{real}^{\vee}$ be the sum of positive real coroots and $2\rho_{im}^{\vee}$ be the sum of positive imaginary coroots for some choice of positive coroots.

Define R_{cx} to be the set of all roots orthogonal to both $2\rho_{real}^{\vee}$ and $2\rho_{im}^{\vee}$.

We observe that R_{cx} is θ-stable because $\theta(2\rho_{real}^{\vee}) = -2\rho_{real}^{\vee}$ and $\theta(2\rho_{im}^{\vee}) = 2\rho_{im}^{\vee}$.

LEMMA 3.1. *The root system R_{cx} is the disjoint union $R_L \coprod R_R$ of two factors interchanged by θ.*

R_{cx} is usually a small subset of the complex roots.

EXAMPLE 3.2. Suppose $G_{\mathbb{C}} = E_8$; we will realize the root system inside the lattice $\mathbb{Z}^8 \cup [\mathbb{Z}^8 + (1/2, \ldots, 1/2)]$. There is a Cartan subgroup for which θ is given by

$$\theta(x_1, x_2, x_3, x_4, x_5, x_6, x_7, x_8) = (-x_1, -x_2, -x_3, -x_4, x_5, x_6, x_7, x_8)$$

$$R_{real} = \{(e_i \pm e_j) | 1 \leq i, j \leq 4\} \text{ of type } D_4$$

$$R_{im} = \{(\pm e_i \pm e_j) | 5 \leq i, j \leq 8\} \text{ of type } D_4$$

$$2\rho^\vee_{real} = (6, 4, 2, 0, 0, 0, 0, 0)$$

$$2\rho^\vee_{im} = (0, 0, 0, 0, 6, 4, 2, 0)$$

$$R_{cx} = (0, 0, 0, \pm 1, 0, 0, 0, \pm 1) \cup (\epsilon_1/2, -\epsilon_1/2, -\epsilon_1/2, \epsilon_2/2, \epsilon_3/2, -\epsilon_3/2, -\epsilon_3/2, \epsilon_4)$$

Here $\epsilon_i = \pm 1$, and the product of the four signs ϵ_i must be 1. Therefore there are twelve roots in R_{cx}; it is easy to check that

$$R_{cx} \simeq A_2 \times A_2,$$

with the two copies of A_2 interchanged by θ.

Here is an ATLAS session producing the above data.

```
[eastman-three-eighty-one:~] anoel% atlas
This is the Atlas of Reductive Lie Groups Software Package version 0.2.4.
Enter "help" if you need assistance.

empty: cartan
Lie type: E8
enter inner class(es): s
(weak) real forms are:
0: e8
1: e8(e7.su(2))
2: e8(R)
enter your choice: 1
Name an output file (hit return for stdout):

Cartan #4:
split: 2; compact: 2; complex: 2
twisted involution orbit size: 3150
imaginary root system: D4
real root system: D4
complex factor: A2
real:
```

We see that the above involution corresponds to Cartan #4.

The following theorem describe the real points of the complex Weyl group.

THEOREM 3.3. *$W(G_{\mathbb{C}}, H_{\mathbb{C}})^\theta \simeq [W(R_{real}) \times W(R_{im})] \rtimes W(R_{cx})^\theta$ and $W(R_L) \simeq W(R_{cx})^\theta$.*

In the case of the above example:

$$W(G_{\mathbb{C}}, H_{\mathbb{C}})^{\theta} \simeq [W(D_4)\times W(D_4)]\rtimes W(A_2) \simeq [(S_4\ltimes(\mathbb{Z}/2\mathbb{Z})^3)\times(S_4\ltimes(\mathbb{Z}/2\mathbb{Z})^3)]\rtimes S_3.$$

To obtain the above information from the above Atlas session one uses the command `realweyl` as follows:

```
real: type
Lie type: E8
enter inner class(es): s
main: realweyl
(weak) real forms are:
0: e8
1: e8(e7.su(2))
2: e8(R)
enter your choice: 1
cartan class (one of 0,1,2,3,4): 4
Name an output file (hit return for stdout):

real weyl group is W^C.((A.W_ic) x W^R), where:
W^C is isomorphic to a Weyl group of type A2
A is trivial
W_ic is a Weyl group of type D4
W^R is a Weyl group of type D4

generators for W^C:
454787
27
generators for W_ic:
45654
67876
3142345423143
245676542
generators for W^R:
345676543
54234567876542345
234565423
1
real:
```

Define $W_{\mathbb{R}}(R_{im}) = W(R_{im}) \cap W(G, H)$. The next theorem gives a description of the Weyl group of H in G.

THEOREM 3.4. $W(G, H) \simeq [W(R_{real}) \times W_{\mathbb{R}}(R_{im})] \rtimes W(R_{cx})^{\theta}$

Applying the previous theorem to our example we see that:

$$W(E_8, H) \simeq [W(D_4) \times W(D_4)] \rtimes W(A_2)$$

$$W(E_8, H) \simeq [[S_4 \ltimes (\mathbb{Z}/2\mathbb{Z})^3] \times [S_4 \ltimes (\mathbb{Z}/2\mathbb{Z})^3]] \rtimes S_3.$$

Define $W_{grad}(R_{im})$ to be the normalizer of $W(R_{cpt})$ in $W(R_{im})$. Then we have

$$W(R_{cpt}) \subseteq W_{\mathbb{R}}(R_{im}) \subseteq W_{grad}(R_{im}) \subseteq W(R_{im}).$$

The Atlas software writes W_{ic} for the Weyl group $W(R_{cpt})$ of the compact imaginary roots.

In our example, all four of these groups are equal, because H is the maximally split Cartan subgroup of G.

In general $W_{grad}(R_{im})$ and $W_{\mathbb{R}}(R_{im})$ are both of the form $W(R_{cpt}) \rtimes (\mathbb{Z}/2\mathbb{Z})^n$.

Define R_T to be the set of roots orthogonal to $2\rho^{\vee}_{real}$. Observe that R_{im} and R_{cx} are subsets of R_T. Furthermore, R_T is θ-stable, contains no real roots and defines a Levi factor L of a real parabolic subgroup of G in which H is fundamental. In fact

$$W(G,H) \simeq W(R_{real}) \rtimes W(L,H).$$

Hence understanding the real Weyl group reduces to the case of the fundamental Cartan $\mathfrak{h} = \mathfrak{t} + \mathfrak{a} \subset \mathfrak{l}$. In this setting $R_{T|_{\mathfrak{t}}}$ is a possibly non-reduced root system and

$$W(L_{\mathbb{C}}, H_{\mathbb{C}})^{\theta} \simeq W(R_{T|_{\mathfrak{t}}}).$$

In our example $L = E_6$. The fundamental Cartan for L has a 4-dimensional compact part, and the restricted root system is $R_T|_{\mathfrak{t}} \simeq F_4$. It turns out that $L \cap K$ is of type F_4, so

$W(L_{\mathbb{C}}, H_{\mathbb{C}})^{\theta} = W(F_4) = W(L \cap K, T).$

This gives an isomorphism

$W(E_8, H) \simeq W(D_4) \rtimes W(F_4).$

It is interesting to reconcile this isomorphism with our earlier calculation

$W(E_8, H) \simeq [W(D_4) \times W(D_4)] \rtimes W(A_2).$

4. Parametrizing Representations of K

Let G be the real points of a complex connected reductive algebraic group $G_{\mathbb{C}}$. Let K be a maximal compact subgroup of G. We parametrize the set $\hat{K}$ of unitary irreducible representations of K. The goal is to describe an algorithm for such a parametrization and to implement it as a package of the *Atlas of Lie groups and representations.*

Let $G_{\mathbb{C}}$ be a complex connected reductive algebraic group and G the set of real points of $G_{\mathbb{C}}$. Let θ be the Cartan involution of G which extends to an involution of $G_{\mathbb{C}}$. We denote by K a maximal compact subgroup of G. Then $G^{\theta}_{\mathbb{C}} = K_{\mathbb{C}}$ the complexification of K. We identify G with a root datum $(X^*, \Delta^+, X_*, (\Delta^+)^{\vee})$. So we would like to describe $\hat{K}$ in term of X^*.

Let H be a θ-stable Cartan subgroup of G and $\Delta(\mathfrak{g}_{\mathbb{C}}, \mathfrak{h}_{c})$ the corresponding root system of $\mathfrak{g}_{\mathbb{C}} = \mathrm{Lie}(G_{\mathbb{C}})$. Δ_{im} and Δ_{re} will denote the sets of imaginary and real roots in $\Delta(\mathfrak{g}_{\mathbb{C}}, \mathfrak{h}_{c})$ respectively. Then $X^*(H_{\mathbb{C}})$ the character lattice of $H_{\mathbb{C}}$ is isomorphic to X^*. Finally, let $T = K \cap H$ a compact, possibly disconnected torus. We have the following lemma:

LEMMA 4.1. *The set of characters of T is isomorphic to* $\frac{X^*(H_\mathbb{C})}{(1-\theta)X^*(H_\mathbb{C})}$.

Let ρ be half the sum of positive roots in $\Delta(\mathfrak{g}_\mathbb{C}, \mathfrak{h}_\mathbb{C})$ and fix

$$\lambda \in \frac{X^*(H_\mathbb{C}) + \rho}{(1-\theta)X^*(H_\mathbb{C})}$$

We want $I(H, \Delta^+_{im}, \Delta^+_{re}, \lambda)$ to correspond to a virtual representation of G restricted to K. Consider a discrete series representation of G restricted to K for example. The main idea is to describe irreducible representations of K as lowest K-types.

Let $2\rho^\vee_\mathbb{R}$ be the sum of positive real coroots. Define

$$\Delta_T = \{\text{roots } \perp 2\rho^\vee_\mathbb{R}\}.$$

Then Δ_T is a θ-stable root system corresponding to a real Levi subgroup L of G with H fundamental in L. Fix Δ^+_T containing Δ^+_{im} and consider the set

$$\mathfrak{L} = \{\lambda \in \frac{X^*(H_\mathbb{C}) + \rho}{(1-\theta)X^*(H_\mathbb{C})}\}$$

such that

(1) λ is weakly dominant for Δ^+_T

(2) if α is a simple imaginary root and $\langle \lambda, \alpha^\vee \rangle = 0$ then α is non-compact.

(3) if β is a simple real root then $\langle \lambda, \beta^\vee \rangle$ is odd.

(1) ensures that $I(H, \Delta^+_{im}, \Delta^+_{re}, \lambda)$ is a standard limit representation of G restricted to K.

(2) ensures that $I(H, \Delta^+_{im}, \Delta^+_{re}, \lambda)$ is non-zero.

(3) ensures that $I(H, \Delta^+_{im}, \Delta^+_{re}, \lambda)$ cannot be written using a more compact Cartan subgroup and Hecht-Schmid identities.

Given that λ is defined modulo $(1-\theta)X^*(H_\mathbb{C})$ to see that property (3) is well-defined one only needs to consider that for $\gamma \in X^*(H_\mathbb{C})$,

$$\langle \gamma - \theta\gamma, \beta^\vee \rangle = \langle \gamma, \beta^\vee - \theta\beta^\vee \rangle = \langle \gamma, 2\beta^\vee \rangle$$

which is even.

The main theorem is:

THEOREM 4.2. *If $\lambda \in \mathfrak{L}$ then $I(H, \Delta^+_{im}, \Delta^+_{re}, \lambda)$ has a unique lowest K-type $\mu(H, \Delta^+_{im}, \Delta^+_{re}, \lambda)$. Hence*

$$\hat{K} = \coprod_{[H \text{ mod conjugation by } K]} \ \coprod_{[\Delta^+_{im} \text{ mod conjugation by } W(G,H)]} \mu(H, \Delta^+_{im}, \Delta^+_{re}, \lambda).$$

To see why conjugation under Δ_{re} does not interfere with this parametrization one has to observe that if $\lambda \simeq \lambda'$ for λ and $\lambda' \in \mathfrak{L}$ then for $\beta \in \Delta_{re}$

$$\lambda' = s_\beta(\lambda) - [\rho_{\mathbb{R}} - s_\beta(\rho_{\mathbb{R}})] = \lambda - (\langle \lambda, \beta^\vee \rangle + 1)\beta = \lambda - 2m\beta \text{ whith } m \in \mathbb{Z}$$

But $2m\beta = m\beta - \theta(m\beta) = m(1-\theta)\beta$.

5. Software for Computing Standard Representations

Let $G_{\mathbb{R}}$ be the real points of a complex connected reductive algebraic group G. Let $K_{\mathbb{R}}$ be a maximal compact subgroup of $G_{\mathbb{R}}$. We describe an algorithm for computing restrictions of standard representations of $G_{\mathbb{R}}$ to $K_{\mathbb{R}}$. We are currently implementing the algorithm as a package of the *Atlas of Lie Groups and Representations* software developed by Fokko du Cloux.

Let θ be a Cartan involution for $\mathfrak{g}_{\mathbb{R}} = Lie(G_{\mathbb{R}})$ then θ extends to an involution of $\mathfrak{g} = Lie(G)$ giving $\mathfrak{g} = \mathfrak{k} + \mathfrak{p}$, the usual Cartan decomposition with $+1$-eigenspace $\mathfrak{k}$ and -1 eigenspace $\mathfrak{p}$. Furthermore $Lie(K) = \mathfrak{k}$. Let B be a Borel subgroup of G and H a Cartan subgroup of G such that $H \subseteq B$. Then K acts on the flag variety G/B. Let H_1 be a θ-stable Cartan subgroup of G defined by the positive root system Δ_1^+ with B_1, the Borel subgroup defined by Δ_1^+. The orbit $K.B_1 = \mathfrak{O}_1$ in G/B determines H_1 up to K-conjugacy. The underlying philosophy is that irreducible $G_{\mathbb{R}}$-representations are to be understood via the algebraic structure of their $(\mathfrak{g}, K_{\mathbb{R}})$ modules.

In Atlas K-orbits on G/B correspond to

$$\coprod_{\text{disjoint union over } K\text{-classes of } \theta\text{-stable } H_i} W(G, H_i)/W(K, H_i).$$

A continued standard representation $(\Delta_{1,im}^+, \lambda_1)$ is given by an orbit $K.B_1$ and a Harish-Chandra module $(Lie(H_1), H_1 \cap K)$.

THEOREM 5.1. *(Hecht, Milicic, Schmid, Wolf)*
(i) Continued standard representations depend only on the imaginary positive roots $\Delta_{1,im}^+$.
(2) If λ_1 *is positive on* $\Delta_{1,im}^+$ *then* $(\Delta_{1,im}^+, \lambda_1)$ *is an actual standard representation.*
(3) Restriction to K *depends only on* $\Delta_{1,im}^+$ *and* $\lambda_1|_{H_1 \cap K}$.

5.1. Example. Let $G_{\mathbb{R}} = U(2,1)$ and $K_{\mathbb{R}} = U(2) \times U(1)$. Then $G = GL(3, \mathbb{C})$ and $K = GL(2, \mathbb{C}) \times GL(1, \mathbb{C})$.

There are two K-conjugacy classes of θ-stable Cartan subgroups:

$H_1 =$ diagonal matrices that is $(\mathbb{C}^*)^3$

and

$$H_2 = \begin{pmatrix} 1 & 0 & 0 \\ 0 & \cosh t & \sinh t \\ 0 & \sinh t & \cosh t \end{pmatrix} \cdot \begin{pmatrix} e^\phi & 0 & 0 \\ 0 & e^\gamma & 0 \\ 0 & 0 & e^\gamma \end{pmatrix}$$

In the H_1 case $\Delta_1^+ = \Delta^+ - 1, im$ and $W(G, H_1) = S_3$. Let $\rho = (1, 0, -1)$. If we want to compute standard representations with infinitesimal character ρ then each λ will correspond to a permutation of ρ with the first two coordinates in decreasing order.

So λ_1 takes the following values; $(1, 0, -1)$, $(1, -1, 0)$ and $(0, -1, 1)$ accounting for 3 discrete series of $U(2, 1)$.

In the case of H_2 there are no imaginary roots and $[W(K, H_{2,\mathbb{R})} = S_1 \times S_2]$. Here λ_2 takes the following values; $(1, 0, -1)$, $(0, 1, -1)$ and $(-1, 1, 0)$. When restricted to K we obtain the three principal series of $U(2, 1)$.

5.2. How to describe restriction to K. First we will work on the fundamental Cartan in order to obtain a simpler formula. We note that $K.B_1 \simeq K/(B_1 \cap K)$ which is a complete flag variety for K and up to a ρ-shift λ_1 corresponds to a line bundle $\mathfrak{L}_{\lambda_1}$ on $K/(B_1 \cap K)$. This is data for discrete series of $G_{\mathbb{R}}$. The lowest K-type is $H^{top}(K/(B_1 \cap K), \mathfrak{L}_{\lambda_1})$ which corresponds to an irreducible representation of K with highest weight λ_1 (up to a $\rho_G - 2\rho_K$)-shift. Here $top = \dim_{\mathbb{C}} K/(B_1 \cap K)$

Ideas from D-module theory require that one adds formal derivatives away from K-orbits. The final formula is:

$$(\Delta_{im}^+, \lambda_1)|_k = \sum_{m \geq 0} \sum_i (-1)^i H^{top-i}(K/(B_1 \cap k), \mathfrak{L}\lambda_1 \otimes S^m(\mathfrak{n} \cap p)).$$

Here $\mathfrak{b}_1 = \mathfrak{h}_1 \oplus \mathfrak{n}$ with $Lie(B_1) = \mathfrak{b}_1$. (The lowest K-type is buried in the above formula for $m = 0$.)

To extract the lowest K-type one has to invert $\sum_{m \geq 0} S^m(\mathfrak{n} \cap p)$ to get $\sum_{j=0}^{\dim(\mathfrak{n} \cap \mathfrak{p})} (-1)^j \wedge^j (\mathfrak{n} \cap \mathfrak{p})$.

Use Koszul Theorem here to set

$$\sum_{j=0}^{\dim(\mathfrak{n} \cap p)} \sum_{m \geq 0} (-1)^j (\wedge^j \otimes S^m)(\mathfrak{n} \cap p) = \mathbb{C}.$$

Finally we obtained,

$$H^{top}(K/(B_1 \cap k), \mathfrak{L}\lambda_1) = \sum_{j=0}^{\dim(\mathfrak{n} \cap p)} (-1)^j \text{standardrep } (\lambda_1 \otimes \wedge^j (\mathfrak{n} \cap p))|_K.$$

5.3. Formula for general Cartan subgroups. Let H_i be a Cartan subgroup of G and fix a set of positive imaginary roots $\Delta_{i,im}^+$. We can extend $\Delta_{i,im}^+$ to a positive root system Δ_i^+ which is as θ-stable as possible, that is if α is positive then either α is real, $\theta(\alpha) = -\alpha$ or α is not real and $\theta(\alpha)$ is positive. This gives some θ-stable parabolic subalgebra $\mathfrak{q}_i = \mathfrak{l}_i \oplus \mathfrak{u}_i$ such that the roots in $\mathfrak{l}_i$ are all real and the ones in $\mathfrak{u}_i$ are not real. Let $\mathfrak{b}_i$ be the Borel subalgebra corresponding to Δ_i^+. Let Q be the unique conjugate of Q_i containing B. Denote by $\mathfrak{q} = \mathfrak{l} \oplus \mathfrak{u}$ the parabolic subalgebra of $\mathfrak{g}$ associated to Q. Then the natural map $G/B \to G/Q$

carries $K.\mathfrak{b}_i \to K.\mathfrak{q}_i \simeq K/(Q_i \cap K)$, a closed orbit in G/Q. This says that the fiber over $\mathfrak{q}_i$ is $(L_i \cap K)/(H_i \cap K)$ which is open (orbit) in $L/(L \cap B)$.

The conclusion is that K-orbit of $\mathfrak{b}_i$ is a fiber bundle over closed $K/(Q_i \cap K)$. There is a standard representation of $L_{\mathbb{R}}$ related to the fiber $(L_i \cap K)/(H_i \cap K)$. It is the sections of the bundle defined by a character $\lambda_i|_{(H_i \cap K)}$:

$$Ind_{H_i \cap K}^{L_i \cap K} \lambda_i|_{(H_i \cap K)} = \bigoplus_{\tau\,:\,\text{irreducible rep of } L_i \cap K} mult(\lambda_i|_{(H_i \cap K)} \text{ in } \tau|_{(H_i \cap K)})$$

5.4. Example. Let $G_{_\mathbb{R}} = GL(3, \mathbb{R})$ and $K_{_\mathbb{R}} = O(3, \mathbb{R})$ then $G = GL(3, \mathbb{C})$ and $K = O(3, \mathbb{C})$. G/B is the variety of flags in $\mathbb{C}^3$.

The closed orbits of K consist of flags $(L \subset L^{\perp})$ where L is a line of length zero . For example a line generated by $(1, i, 0)$ included in the plane generated by $(1, i, 0)$ and $(0, 0, i)$. According to Wiit's theorem any two lines of length zero are conjugate by K.

The open orbits of K consist of flags $(L \subset P)$ such that L and $P^{\perp}$ are both of non zero length. An example would be the line generated by $(1, 0, 0)$ in plane P generated by $(1, 0, 0)$ and $(0, 1, 0)$. A counter example would be the line generated by $(1, 0, 0)$ in plane P generated by $(1, 0, 0)$ and $(0, 1, i)$ since in this case $P^{\perp}$ is generated by $(0, 1, i)$ which is of length zero. (Witt's theorem says such pairs (line, plane) are a single orbit.

The stabilizer of $(1, 0, 0) \subset \langle (1, 0, 0), (0, 1, 0) \rangle$ is the set of upper triangular matrices in K that is $(\pm 1, \pm 1, \pm 1)$.

The complete flag in $\mathbb{C}^3$ contains $O(3, \mathbb{C})/O(1, \mathbb{C})^3$ (open)
λ is a character of $O(1, \mathbb{C})^3 = \{\epsilon_1, \epsilon_2, \epsilon_3\}$
Sections of this λ-bundle are
$Ind_{O(1,\mathbb{C})^3}^{O(3,\mathbb{C})} \lambda|_{(H_i \cap K)}$ = sum of all $O(3, \mathbb{C})$ irreducible representations τ with multiplicities = mult of λ in $\tau|_{O(1,\mathbb{C})^3}$

5.5. Zuckerman Formula. Trivial rep of $K =$
$\sum_{KGBorbits:\mathfrak{O}} (-1)^{\text{codim } \mathfrak{O}} \text{ stdrep}(\mathfrak{O}, triv\lambda)|_K$
One of the terms in the sum is the standard representation for this open orbit restricted to K + terms for lower dimensional orbits.

List of all orbits for $O(3, \mathbb{C})$ on complete flags:
dim 3 open non-zero length $L \subset P$ with $P^{\perp}$ non-zero length
dim 2 zero length $L \subset P \neq L^{\perp}$
dim 2 non-zero length $L \subset P$ with $P^{\perp}$ zero length
dim 1 closed zero length $L \subset L^{\perp}$

Standard representations data includes λ_i, character of $H_i \cap K$. So we have this algebraic bundle over K orbits and we are interested in sections. Since the L_i orbit is open in the fiber one can differentiate in those directions (sections of bundle on K orbit). We need to add derivatives that are transverse corresponding to $\mathfrak{u}_i \cap \mathfrak{p}$. (This is away from K-orbit directions in which you can't differentiate)

THEOREM 5.2. *$StdReps|_K = \sum_p (-1)^p H^{top-p}(K/(Q_i \cap K)$, stdrep for $L_i|_{L_i \cap K} \otimes S(\mathfrak{u}_i \cap \mathfrak{p}))$ with $top = \dim_{\mathbb{C}} K/(Q_i \cap K)$.*

Hiding inside of the above formula is the lowest K-ype

$$H^{top}(K/(Q_i \cap K), \text{lowest } (L_i \cap K) - \text{type of strep for} L_i) \otimes S^0(\mathfrak{u}_i \cap \mathfrak{p}).$$

How do you write lowest K-type as combination of standard representations?

Use Zuckerman formula:

Lowest $L_i \cap K$-type $= \sum_{\mathfrak{O}} (-1)^{codim \mathfrak{O}} \text{stdrep} \,|_{L_i \cap K}$

where $\mathfrak{O}$ is an orbit of $L_i \cap K$ on $L_i/(L_i \cap B)$

Lowest K-type $= \sum_{\mathfrak{O}} (-1)^{codim \mathfrak{O}} H^{top}(K/(Q_i \cap K), \text{stdrep for} L_i|_{L_i \cap K}, \mathfrak{O})$

We need to put the transverse derivatives in. Using Koszul identity

$$S(\mathfrak{u} \cap \mathfrak{p}) \otimes \sum_{j=0}^{\dim(\mathfrak{u} \cap p)} (-1)^j (\wedge^j(\mathfrak{u} \cap p)) = \mathbb{C}.$$

Lowest K-type=

$$\sum_{j=0}^{\dim(\mathfrak{u} \cap p)} \sum_{\mathfrak{O}} (-1)^j H^{top}(K/(Q_i \cap K), (\text{stdrep for } L_i|_{L_i \cap K}, \mathfrak{O}, \lambda_i) \otimes \wedge^j(\mathfrak{u} \cap p) \otimes S(\mathfrak{u} \cap \mathfrak{p}))$$

$$= \sum_{\text{subset of roots of size j in } \mathfrak{u} \cap \mathfrak{p}} \sum_{\mathfrak{O}} (-1)^{codim \mathfrak{O}} \text{stdrep for } G_{\mathbb{R}}.$$

To compute the Zuckerman terms one proceeds as follows:

1. For each Cartan subgroup H_i construct a θ-stable parabolic subalagebra $\mathfrak{q}_i = \mathfrak{l}_i \oplus \mathfrak{u}_i$ such that L_i is split (all roots are real).

2. Call (cartan (L_i)) to obtain normalized involutions $\{\theta_i^j\}$. Zuckerman formula for L_i is indexed by the KGB orbits for $L_i \cap K$ on $L_i/(B \cap L_i)$. These orbits are indexed by Cartan subgroups $\{\theta_i^j\}$ and correspond to

$$\coprod_{j \leftrightarrow H_i^j} W(L_i, H_i)/W(L_i \cap K, H_i^j).$$

What emerge are cosets $w.W(L_i, H_i)/W(L_i \cap K, H_i^j)$ corresponding to the characters $(w.\rho_{L_i} + \rho_{L_i})|_{(H_i^j \cap K)}$ contributing to the Zuckerman formula with some sign $(-1)^{\text{length(KGBELT)}}$.

3. For each j set a pair (m, μ) with $m \in \mathbb{Z}$ and μ a character of $H_i^j \cap K$.

4. Compute $\sum_{[j:\#\text{ of Cartans in } L_i]} \sum_{(m,\mu)} m.\text{stdrep}(H_i^j, \lambda_i^j + \mu)$

In general H_i^J will be more compact than H_i because of Cayley transform in real roots in L_i. Roughly $H_i \cap K \subseteq H_i^j \cap K$. So $\lambda_i \to \lambda_i^j$. The roots of $\alpha_1 \dots \alpha_l$ in H_i^j are orthogonal to the real roots in L_i and $H_i^j = (H_i \cap K).(SO_2)^l$. We define $\lambda_i^j = \lambda_i$ and trivial on the SO_2 factors.

For the outer sum list all the roots of $H_i^j \cap K$ in $\mathfrak{u} \cap \mathfrak{p}$.

6. Fokko du Cloux December 20, 1954 - November 10, 2006

Fokko du Cloux was born in Holland. The name du Cloux seems to come from an old French Huguenot protestant family who moved to Holland probably around 1685 after the revocation of the Edict of Nantes by Louis XIV in order to avoid persecution from the Roman Catholic pope. When he was young he lived in Spain and attended the Lycée Français. He was a student of Alain Guichardet at the École Polytechnique at the same time as Patrick Delorme, who is now in Marseille, and stayed there as a CNRS researcher until he became a Professor at the Université de Lyon in 1991.

For ten years he developed *Coxeter* which he describes as a computer program for the study of combinatorial aspects of Coxeter group theory, particularly those related to the Bruhat ordering and Kazhdan-Lusztig polynomials. It is written in C++, and should run on the main types of workstation under linux or unix. The current version is now version 3.0. Visit
http://math.univ-lyon1.fr/~ducloux/coxeter/coxeter3/english/coxeter3_e.html
for more information about Coxeter.

From 2003 to 2006 he was the main developer of the Atlas of Lie Groups and Representations software.

6.1. From the American Mathematical Society: Fokko du Cloux died on November 10 at the age of 52. He was born December 20, 1954, in Rheden in the Netherlands. Du Cloux grew up in Spain, and was an undergraduate at École Polytechnique in Paris, graduating in 1978. He completed a Docteur d'État there in 1984, under Alain Guichardet. From 1985 to 1991 he was a Chargé de Recherche of the CNRS at École Polytechnique, and from 1991 a professor at Université Lyon-I. Du Cloux combined deep insight into the mathematics of Coxeter groups with tremendous skill as a programmer. His program Coxeter sets the standard for computations in these groups. During the last four years of his life he began work on an equally powerful and general program Atlas for the structure and representation theory of real reductive groups. He was an AMS member since 1984.

6.2. From Jeffrey Adams: I met Fokko in 1987 in Berkeley, although only barely. I think the second time I saw him was in Montreal, in June 2002 at the conference Bill Casselman helped organize on computers and mathematics. [1]

Fokko is an inspiration to me. He both accepted his illness and fought it at the same time. I never heard him complain, and his terrible situation never got in the way of his love of Mathematics. The fact that he worked productively and happily with David, Marc van Leeuwen and myself that Fall is a marvel I will carry with me always.

The subject of representations of real reductive groups is a notoriously specialized one. It has been a source of some frustration to me over the years that so few people appreciate the subject. I am very excited by the prospect that the software Fokko has written will make this work accessible to a much wider audience.

I am equally excited by what the software has already done for the experts. I have learned a lot about the mathematics by looking at examples produced by

[1]Workshop on Computational Lie Theory, Centre de recherches mathématiques Montreal, QC Canada May 27 - June 7, 2002.

the software (not to mention what I've learned working with Fokko along the way). The fact that David Vogan says much the same thing is a true testament to the value of the work that Fokko has done.

Members of the Atlas of Lie groups and representations

Jeffrey Adams	Dan Barbasch	Birne Binegar	Bill Casselman
Dan Ciubotaru	Fokko du Cloux	Scott Crofts	Tatiana Howard
Monty Mcgovern	Alfred Noël	Alessandra Pantano	Annegret Paul
Patrick Polo	Siddhartha Sahi	Susana Salamanca	John Stembridge
Peter Trapa	Marc van Leeuwen	David Vogan	Wai-Ling Yee
Jiu-Kang Yu	Gregg Zuckerman		

References

[F] F. du Cloux, *Computing Kazhhan-Lusztig Polynomials for Arbitrary Coxeter Groups*, Experimental Mathematics, **11** (3), (2001), 371–381.

[HC] Harish-Chandra, *Invariant eigendistributions on a semisimple Lie group*, Trans. Amer. Math. Soc., **119** (1965) 457–508.

[K] A. W. Knapp, *Representation theory of semisimple groups. An overview based on examples* Princeton Mathematical Series, **36**. Princeton University Press, Princeton, NJ, 1986

[L] R. P. Langlands, *On the classification of irreducible representations of real algebraic groups* Representation theory and harmonic analysis on semisimple Lie groups, 101–170, Math. Surveys Monogr., **31**, Amer. Math. Soc., Providence, RI, 1989

[V] D. Vogan, *The Character Table for E_8*, Notices of the AMS, **54**, (9), (2007), 1022–1034.

[VL] M. A. A. Van Leeuwen, A. M. Cohen , B. Lisser *LiE Apackage for Lie Group Computations*, Computer Algebra Nederland, Amsterdam The Netherlands 1992.

Department of Mathematics, University of Massachusetts, 100 Morrissey Boulevard, Boston, MA 02125-3393

E-mail address: anoel@math.umb.edu

Contemporary Mathematics
Volume **467**, 2008

Maximal Tori of Reductive Centralizers of Nilpotents in Exceptional Complex Symmetric Spaces: A Computational Approach

Alfred G. Noël

ABSTRACT. The maximal tori and triples described in this paper arise naturally in the study of nilpotent orbits of Lie groups and play an important rôle in several problems such as: classification of nilpotent orbits of real Lie groups [**No**], description of admissible nilpotent orbits of real Lie groups [**No1**], [**No2**], [**O**], [**Sch**], classification of spherical nilpotent orbits [**K2**], [**K3**], determination of component groups of centralizers of nilpotents in symmetric spaces [**K1**], [**K-N**].

Introduction

The main goal of this paper is to make available to researchers a set of data that could make it easier to test conjectures arising from questions about maximal tori in the reductive centralizer of nilpotent elements. As Computer Algebra Systems and computer hardware become more and more powerful we expect them to be a very important part of the pure mathematician's toolbox.

Explicit information about reductive centralizers of nilpotents is hard to obtain. In many cases, questions about representations of such centralizers are answered via data available from their maximal tori. In this paper we give an explicit description of such tori and their associated normal triples.

There are two fundamental aspects of scientific computing. The first one is algorithm design and the other is algorithm implementation. Here we use non trivial Lie theory results to develop our algorithm. The implementation is done using two software packages *LiE* ([**VL**]) and Mathematica ([**W**]) in addition to our own programs. Readers who want to know more about implementation details and software engineering issues should consult [**No3**].

In order to proceed we need some definitions. Let $\mathfrak{g}$ be a real semisimple Lie algebra with adjoint group G and $\mathfrak{g}_{\mathbb{C}}$ its complexification. Also let $\mathfrak{g} = \mathfrak{k} \oplus \mathfrak{p}$ be a Cartan

1991 *Mathematics Subject Classification.* 17B05,17B10,17B20,22E30.
Key words and phrases. Maximal tori, nilpotent orbits, reductive Lie algebras.
The author was partially supported by NSF grant #DMS 0554278

decomposition of $\mathfrak{g}$. Finally, let θ be the corresponding Cartan involution of $\mathfrak{g}$ and σ be the conjugation of $\mathfrak{g}_{\mathbb{C}}$ with regard to $\mathfrak{g}$. Then $\mathfrak{g}_{\mathbb{C}} = \mathfrak{k}_{\mathbb{C}} \oplus \mathfrak{p}_{\mathbb{C}}$ where $\mathfrak{k}_{\mathbb{C}}$ and $\mathfrak{p}_{\mathbb{C}}$ are obtained by complexifying $\mathfrak{k}$ and $\mathfrak{p}$ respectively. Denote by $K_{\mathbb{C}}$ the connected subgroup of the adjoint group $G_{\mathbb{C}}$ of $\mathfrak{g}_{\mathbb{C}}$, with Lie algebra $\mathfrak{k}_{\mathbb{C}}$.

The Kostant-Sekiguchi correspondence

The Kostant-Sekiguchi correspondence is a bijection between nilpotent orbits of G in $\mathfrak{g}$ and nilpotent orbits of $K_{\mathbb{C}}$ on $\mathfrak{p}_{\mathbb{C}}$. Thus, it allows us to study certain questions about real nilpotent orbits by looking at nilpotent orbits of $K_{\mathbb{C}}$ on the symmetric space $\mathfrak{p}_{\mathbb{C}}$. Here is a brief description of the correspondence.

A triple (x, e, f) in $\mathfrak{g}_{\mathbb{C}}$ is called a standard triple if $[x, e] = 2e$, $[x, f] = -2f$ and $[e, f] = x$. If $x \in \mathfrak{k}_{\mathbb{C}}$, e and $f \in \mathfrak{p}_{\mathbb{C}}$ then (x, e, f) is a normal triple. It is a result of Kostant and Rallis **[K-R]** that any nilpotent e of $\mathfrak{p}_{\mathbb{C}}$ can be embedded in a standard normal triple (x, e, f). Moreover e is $K_{\mathbb{C}}$-conjugate to a nilpotent e' inside of a normal triple (x', e', f') with $\sigma(e') = f'$ **[Se]**. The triple (x', e', f') will be called a $Kostant - Sekiguchi$ or KS-triple .

Every nilpotent E' in $\mathfrak{g}$ is G-conjugate to the element E of a triple (H, E, F) in $\mathfrak{g}$ with the property that $\theta(H) = -H$ and $\theta(E) = -F$ **[Se]**. Such a triple will be called a KS-triple also.

Define a map c from the set of KS-triples of $\mathfrak{g}$ to the set of normal triples of $\mathfrak{g}_{\mathbb{C}}$ as follows:

$$x = c(H) = i(E - F)$$

$$e = c(E) = \tfrac{1}{2}(H - i(E + F))$$

$$f = c(F) = \tfrac{1}{2}(H + i(E + F))$$

The triple (x, e, f) is called the Cayley transform of (H, E, F). It is easy to verify that the triple (x, e, f) is a KS-triple and that $x \in i\mathfrak{k}$. The Kostant-Sekiguchi correspondence **[Se]** gives a one to one map between the set of G-conjugacy classes of nilpotents in $\mathfrak{g}$ and the $K_{\mathbb{C}}$-conjugacy classes of nilpotents in $\mathfrak{p}_{\mathbb{C}}$. This correspondence sends the zero orbit to the zero orbit and the orbit through the nilpositive element of a KS-triple to the one through the nilpositive element of its Cayley transform.

Let $\mathfrak{k}_{\mathbb{C}}^{(x,e,f)}$ be the centralizer of (x, e, f) in $\mathfrak{k}_{\mathbb{C}}$. Then it is known that $\mathfrak{k}_{\mathbb{C}}^{(x,e,f)}$ is a reductive Lie algebra. Let $\mathfrak{t}$ be a Cartan subalgebra of $\mathfrak{g}_{\mathbb{C}}$ such that $x \in \mathfrak{t}$. Then we want to find a maximal torus $\mathfrak{t}_1$ of $\mathfrak{k}_{\mathbb{C}}^{(x,e,f)}$ such that $\mathfrak{t}_1 \subseteq \mathfrak{t}$.

The reader should be aware that in general $\mathfrak{t}_1 \neq \mathfrak{t}^{(x,e,f)}$. A counterexample can be found in **[No1]**. Furthermore, there is currently no good characterization of such a torus in the literature. And our conversation with several experts led us to believe that such characterization may be quite technical.

Our contribution consists of the solution of the problem for complex non compact exceptional symmetric spaces via the algorithm described below. First we describe three different problems where an explicit knowledge of the structure of such maximal tori of the reductive centralizer of nilpotents plays a pivotal rôle.

Classification of real nilpotent orbits of G on $\mathfrak{g}$

The main result in **[No]** is a classification of the nilpotent orbits of the real adjoint Lie group G on its real Lie algebra $\mathfrak{g}$. The classification is an extension of the Bala-Carter classification for complex Lie groups. We were able to use the Kostant-Sekiguchi correspondence to classify the nilpotent orbits of $K_{\mathbb{C}}$ on $\mathfrak{p}_{\mathbb{C}}$ instead. Maintaining the above notations the main result is:

THEOREM [NOËL]. *There is a 1-1 correspondence between KS-triples (x, e, f) and triples $(l, \mathfrak{q}, \mathfrak{w})$ where l is a (θ, σ)-stable minimal Levi subalgebra of $\mathfrak{g}_{\mathbb{C}}$ containing e, $\mathfrak{q}$ a θ-stable parabolic subalgebra of the derived algebra of l and $\mathfrak{w}$, a certain $L \cap K_{\mathbb{C}}$ sub-module of $u \cap \mathfrak{p}_{\mathbb{C}}$ with $\mathfrak{q} = l \oplus u$. (L is the connected Lie group of $G_{\mathbb{C}}$ with Lie algebra l).*

Proof. See **[No]**.

One of the main components in the proof of the above result is the isolation of a set of nilpotents called *noticed* nilpotents. They are characterized by the fact that the centralizer $\mathfrak{k}_{\mathbb{C}}^{(x,e,f)}$ is trivial. Furthermore any Cartan subalgebra of $\mathfrak{k}_{\mathbb{C}}^{(x,e,f)}$ is a maximal toral of $\mathfrak{k}_{\mathbb{C}}^{e}$ and the next proposition tells us how to find a θ-stable minimal Levi subalgebra of $\mathfrak{g}_{\mathbb{C}}$ containing e.

PROPOSITION [NOËL]. *If l is a minimal (σ, θ)-stable Levi subalgebra of $\mathfrak{g}_{\mathbb{C}}$ containing a nilpotent element e of $\mathfrak{p}_{\mathbb{C}}$ then $l = \mathfrak{g}_{\mathbb{C}}^{\mathfrak{t}}$, where $\mathfrak{t}$ is a maximal toral subalgebra of $\mathfrak{k}_{\mathbb{C}}^{e}$.*

Proof. See **[No]**.

Hence, an explicit knowledge of $\mathfrak{t}$ would make the problem of computing the bijection more tractable.

Classification of admissible real nilpotent orbits of G on $\mathfrak{g}$

The admissible nilpotent orbits of real simple Lie algebras are now classified. For classical Lie algebras the classification was achieved by J. Schwartz **[Sch]** and T. Ohta **[O]** in 1987 and 1991 respectively. The author has, recently, determined such orbits for the real exceptional Lie algebras **[No1]**, **[No2]**. The general approach is the following: the problem of determining admissible nilpotent orbits of a simple real Lie algebra $\mathfrak{g}$ with Lie Group G is translated into that of classifying admissible nilpotent orbits of the complex symmetric space $\mathfrak{p}_{\mathbb{C}}$ attached to $\mathfrak{g}$. This is made possible by the Kostant-Sekiguchi bijection between real G nilpotent orbits on $\mathfrak{g}$ and complex $K_{\mathbb{C}}$-nilpotent orbits on $\mathfrak{p}_{\mathbb{C}}$, where $K_{\mathbb{C}}$ is a connected subgroup of $G_{\mathbb{C}}$, the complexification of G **[Se]**, and by the fact that the bijection preserves admissibility of associated orbits **[Schwartz]**. The study of these orbits is important because they seem to be good canditates for which a general method for quantization, as predicted by the Orbit method, could be established. The reader may consult **[K]**, **[A-K]**, **[D]**, **[V]**,**[V1]** for more information.

Let e be a non zero nilpotent in $\mathfrak{p}_{\mathbb{C}}$. Then $K_{\mathbb{C}}^e$ acts on $\mathfrak{k}_{\mathbb{C}}/\ \mathfrak{k}_{\mathbb{C}}^e$ and $(\mathfrak{k}_{\mathbb{C}}/\mathfrak{k}_{\mathbb{C}}^e)^*$. Define the character δ_e of $K_{\mathbb{C}}^e$ as follows:

$$\delta_e(g) = (det(g|_{\mathfrak{k}_{\mathbb{C}}/\mathfrak{k}_{\mathbb{C}}^e}))^{-1} \qquad g \in K_{\mathbb{C}}^e$$

Using δ_e and the homomorphism $s : \mathbb{C}^\times \to \mathbb{C}^\times$, with $s(z) = z^2$ we obtain the following double cover of $K_{\mathbb{C}}^e$:

$$\tilde{K}_{\mathbb{C}}^e = \{(g, z) \in K_{\mathbb{C}}^e \times \mathbb{C}^\times : \delta_e(g) = z^2\}.$$

The following lemma of Ohta shows the importance of maximal tori of the reductive centralizer in determining admissible nilpotent orbits:

LEMMA (OHTA). *Let $\mathfrak{t}_1$ be a Cartan subalgebra of $\mathfrak{k}_{\mathbb{C}}^{(x,e,f)}$ and T_1 the corresponding connected subgroup of $(K_{\mathbb{C}}^{(x,e,f)})_\circ$, the identity component of $K_{\mathbb{C}}^{(x,e,f)}$. Then e is admissible if and only if there exists a character, χ, of T_1 such that $\delta_e(g) = (\chi(g))^2$ for all $g \in T_1$.*

Proof. See **[O]**.

COROLLARY [OHTA]. *Suppose that $T_1 \simeq (C^\times)^r$. Then $d\delta_e|_{\mathfrak{t}_1}$ is a linear map : $\mathbb{C}^r \longrightarrow \mathbb{C}$. From Ohta's lemma e is admissible if and only if each coefficient of z_i in the linear combination of $(z_1, z_2, \ldots z_r)$ is an even integer.*

Proof. Obvious from Lemma 2.

Classification of spherical nilpotent orbits of $K_{\mathbb{C}}$ on $\mathfrak{p}_{\mathbb{C}}$

Assume that $G_{\mathbb{C}}$ is connected. Let X be an algebraic variety on which $G_{\mathbb{C}}$ acts. We say that X is spherical for $G_{\mathbb{C}}$ if there exists a Borel subgroup of $G_{\mathbb{C}}$ that has an open dense orbit in X. Some important spherical varieties consist of nilpotent elements associated to symmetric spaces.

The spherical nilpotent orbits of $G_{\mathbb{C}}$ in $\mathfrak{g}_{\mathbb{C}}$ were determined by Panyushev **[Pa]** and McGovern **[McG]**. The nilpotent orbits of $K_{\mathbb{C}}$ on $\mathfrak{p}_{\mathbb{C}}$ are studied extensively by many authors. The fundamental results are from Kostant and Rallis **[K-R]**. It is known that non-zero minimal nilpotent orbits of $K_{\mathbb{C}}$ on $\mathfrak{p}_{\mathbb{C}}$ are spherical. Moreover the Orbit method attaches interesting unitary representations to such orbits **[A-H-V]**, **[V]**.

Recently, Donald R. King gave a characterization of spherical nilpotent orbits as follows:

THEOREM, [KING]. *A nilpotent orbit $\mathfrak{O}$ of $K_{\mathbb{C}}$ on $\mathfrak{p}_{\mathbb{C}}$ is spherical for $K_{\mathbb{C}}$ if and only if its image $\Omega_{\mathfrak{O}}$ in $\mathfrak{g}$ under the Kostant-Sekiguchi correspondence is multiplicity free as a Hamiltonian K-space.*

Proof. See [**K2**].

DEFINITION. *Let $H \subseteq M$ be reductive subgroups of $G_{\mathbb{C}}$. Then H is spherical in M if for each finite dimensional irreducible representation V of M, a finite dimensional irreducible representation W of H occurs in the restriction of V to H with multiplicity at most one.*

Let $\mathfrak{u}$ denote the sum of the positive eigen eigenspaces of ad_x on $\mathfrak{g}_{\mathbb{C}}$ and set

$$Z = \frac{\mathfrak{u} \cap \mathfrak{k}_{\mathbb{C}}}{(\mathfrak{u} \cap \mathfrak{k}_{\mathbb{C}})^e}$$

It is possible to define a positive definite Hermitian inner product on $\mathfrak{g}_{\mathbb{C}}$ relative to which one has the following decomposition:

$$\mathfrak{k}_{\mathbb{C}}^x = \mathfrak{k}_{\mathbb{C}}^{(x,e,f)} \oplus \mathfrak{m}$$

Furthermore $K_{\mathbb{C}}^{(x,e,f)}$ acts on $\mathfrak{m}$. Let z be such that the dimension of $K_{\mathbb{C}}^{(x,e,f)}.z$ is maxmal in $\mathfrak{m}$. The set of all such z is open and dense in $\mathfrak{m}$. Define $\mathfrak{S}$ to be the stabilizer of z in $K_{\mathbb{C}}^{(x,e,f)}$. Then

COROLLARY [KING]. *Maintaining the above notations $\mathfrak{O}$ is spherical if and only if $K_{\mathbb{C}}^{(x,e,f)}$ is spherical in $K_{\mathbb{C}}^x$ and a Borel subgroup of $\mathfrak{S}$ has an open dense orbit on Z.*

Proof. See [**K2**].

In many cases $K_{\mathbb{C}}^x = H_{\mathbb{C}} \times H_{\mathbb{C}}$ where $H_{\mathbb{C}}$ is semisimple. If z is regular then $H_{\mathbb{C}}.z$ is maximal in $\mathfrak{m}$ and the stabilizer $\mathfrak{S}$ defined above is a maximal torus of $K_{\mathbb{C}}^{(x,e,f)}$ and its corresponding maximal torus in $\mathfrak{k}_{\mathbb{C}}^{(x,ef)}$ is embeded diagonally in $\mathfrak{k}_{\mathbb{C}}^x$.

In general the action of $\mathfrak{S}$ can be understood from the study of that of $K_{\mathbb{C}}^{(x,e,f)}$. In many cases, such information can be recovered from the action of a maximal torus in $\mathfrak{k}_{\mathbb{C}}^{(x,e,f)}$.

Description of Maximal Tori of Exceptional Complex Symmetric Spaces

We hope that the reader is convinced of the importance of the maximal tori that we are about to describe. Next, we explain the technique that we use.

In his classification of $K_{\mathbb{C}}$-nilpotent orbits of $\mathfrak{p}_{\mathbb{C}}$, D. Djoković **[D1]**,**[D2]** labelled each $K_{\mathbb{C}}$-orbit by the values of the simple roots of $\mathfrak{k}_{\mathbb{C}}$ on the neutral element x of the normal triple (x, e, f) associated to the orbit. He also gave a list of minimal regular semisimple subalgebras containing the nilpotent e up to $K_{\mathbb{C}}$-conjugacy and the type of $\mathfrak{k}_{\mathbb{C}}^{(x,e,f)}$. Regular semisimple minimal algebras are used in the Dynkin classification. We refer the reader to **[Dy]**,**[Dy1]** for more information on such algebras.

Using information from Djoković we came up with the following algorithm to compute maximal tori of the reductive centralizer:

Algorithm.

Begin

Let $\mathfrak{g}$ be one of the simple real Lie algebras under consideration.

1. Choose a set of simple roots for $\mathfrak{k}_{\mathbb{C}}$.
2. **For** each $K_{\mathbb{C}}$-nilpotent orbit in $\mathfrak{p}_{\mathbb{C}}$ **do**
    3. Compute the corresponding normal triple (x, e, f)
    4. Compute a maximal torus $\mathfrak{t}_{\mathbb{C}}^1 \subseteq \mathfrak{k}_{\mathbb{C}}^{(x,e,f)}$

EndFor

End.

Implementation of the algorithm.

To implement step 1 we use the Vogan systems given in **[Kn]** for all inner type groups. For the two non inner type real forms of E_6 we use the root systems given in **[D2]** and **[No2]**.

To compute the neutral element associated with the triple (x, e, f) we use the label given by Djoković to solve a very simple system of linear equations which gives the coefficients of x in the Bourbaki roots system of $\mathfrak{g}_{\mathbb{C}}$. The computation of e is more involved. However we were able to use the type of the minimal regular semisimple containing e provided by Djoković and explicit knowledge of the 2-eigen-space of x in $\mathfrak{p}_{\mathbb{C}}$ in order to realize e as a specific linear combinations of root vectors of $\mathfrak{g}_{\mathbb{C}}^{(2)} \cap \mathfrak{p}_{\mathbb{C}}$. This the hardest part of the software engineering due to the fact that the search in $\mathfrak{g}_{\mathbb{C}}^{(2)} \cap \mathfrak{p}_{\mathbb{C}}$ may be very time consuming. Nevertheless we were able to develop enough techniques to successfuly compute the nilpotent in all cases of interest. Hence we have the normal triple (x, e, f) since we can deduce f from e by replacing every root vector X_β in e by $X_{-\beta}$.

The implementation of step 4 is made easier by the fact that we know the type of $K_{\mathbb{C}}^{(x,e,f)}$ from Djoković's tables. Consequentlty were can set up and solve the appropriate system of equations to find a basis for a maximal torus of $\mathfrak{k}_{\mathbb{C}}^{(x,e,f)}$.

Remark: In all the cases that we consider a basis vector $H \in \mathfrak{t}_{\mathbb{C}}^1$ is always an integer linear combination of the vectors $H_{\alpha_1}, \ldots, H_{\alpha_l}$ corresponding to the Bourbaki simple roots $\Delta = \{\alpha_1, \ldots, \alpha_l\}$ of $\mathfrak{g}_{\mathbb{C}}$. We always choose $H = \sum_{i=1}^{l} a_i H_i$ so that the non- zero $a_i' s$ do not have a non trivial common divisor. In other words H is not a multiple of another vector. This is important for certain applications. Determination of admissibility is an example of such applications.

The computations were carried out on an old Power Macintosh G3 with 128 megabytes RAM.

An example

This example illustrates the above implementation on a specific orbit. We will also show how to compute a minimum θ-stable Levi subalgebra of $\mathfrak{g}_{\mathbb{C}}$ containing the nilpotent representing the orbit. This the first step in computing the parametrization defined in [**No**].

Let $\mathfrak{g}$ be EII (or $E_{6(2)}$) a real form of E_6 and let $\Delta = \{\alpha_1, \alpha_2, \ldots, \alpha_6\}$ the Bourbaki simple roots of $\mathfrak{g}_{\mathbb{C}}$. Then $\Delta_k = \{\beta_1, \ldots, \beta_6\}$, where $\beta_1 = \alpha_1$, $\beta_2 = \alpha_3$, $\beta_3 = \alpha_4$, $\beta_4 = \alpha_5$, $\beta_5 = \alpha_6$ and $\beta_6 = \alpha_1 + 2\alpha_2 + 2\alpha_3 + 3\alpha_4 + 2\alpha_5 + \alpha_6$, is a set of simple roots for $\mathfrak{k}_{\mathbb{C}} = \mathfrak{sl}_6(\mathbb{C}) \oplus \mathfrak{sl}_2(\mathbb{C})$. The root system defined by Δ_k is a Vogan system. See Knapp [**Kn**] for more information on such systems.

We consider orbit 6 labeled "00000 4" in Djokovic's classification [**D1**]. We use the given label to compute the neutral element x as follows:

Assume that $x = \sum_{i=1}^{6} \mu_i H_{\alpha_i}$ with $\mu_i \in \mathbb{Z}$. Then we solve the system

$$\beta_i(x) = 0 \quad \text{for } i = 1, 2, 3, 4, 5 \quad \text{and} \quad \beta_6(x) = 4$$

to find $x = 2H_{\alpha_1} + 4H_{\alpha_2} + 4H_{\alpha_3} + 6H_{\alpha_4} + 4H_{\alpha_5} + 2H_{\alpha_6}$.

Next we compute $\mathfrak{g}_{\mathbb{C}}^2 \cap \mathfrak{p}_{\mathbb{C}}$ the 2-eigenspace of x intersected with $\mathfrak{p}_{\mathbb{C}}$ and then search that space in order to find

$$e = \sqrt{2}(X_{\alpha_1+\alpha_2+\alpha_3+2\alpha_4+\alpha_5} + X_{\alpha_2+\alpha_3+\alpha_4+\alpha_5+\alpha_6})$$

We should point out that it is sometimes computationally intensive to find e for the dimension of $\mathfrak{g}_{\mathbb{C}}^2 \cap \mathfrak{p}_{\mathbb{C}}$ may be big.

In this case $\mathfrak{k}_{\mathbb{C}}^{(x,e,f)}$ is of type $3A1+T_1$ (see [**D1**]) and we find that

$$\mathfrak{t}_{\mathbb{C}}^1 = \mathbb{C}H_{\alpha_5} \oplus \mathbb{C}(H_{\alpha_1}+H_{\alpha_3}) \oplus \mathbb{C}(H_{\alpha_4}+H_{\alpha_6}) \oplus \mathbb{C}(H_{\alpha_3}+H_{\alpha_4})$$

is the maximal torus that we want.

To find a minimal θ-stable Levi subalgebra containing e it is enough to compute $l = \mathfrak{g}_{\mathbb{C}}^{\mathfrak{t}_{\mathbb{C}}^1}$, the centralizer of $\mathfrak{t}_{\mathbb{C}}^1$ in $\mathfrak{g}_{\mathbb{C}}$. And easy computations shows that:

$$l = \mathbb{C}X_{\pm(\alpha_1+\alpha_2+\alpha_3+2\alpha_4+\alpha_5)} \oplus \mathbb{C}X_{\pm(\alpha_2+\alpha_3+\alpha_4+\alpha_5+\alpha_6)} \oplus \mathbb{C}X_{\pm(\alpha_1+2\alpha_2+2\alpha_3+3\alpha_4+2\alpha_5+\alpha_6)}.$$

Hence l is of type A_2.

Tables of Results

Remark: Except for type G_2 the above maximal tori were computed for all non even nilpotent orbits ([**No1**],[**No2**]). If the nilpotent e is noticed then the reductive centralizer is trivial. In the case of the zero orbit the reductive centralizer is $\mathfrak{k}_{\mathbb{C}}$ hence any Cartan subalgebra of $\mathfrak{k}_{\mathbb{C}}$ will suffice. Therefore, it is enough to compute maximal tori of $\mathfrak{k}_{\mathbb{C}}^{(x,e,f)}$ for non-trivial, non-noticed, even nilpotent orbits of $K_{\mathbb{C}}$ on $\mathfrak{p}_{\mathbb{C}}$ for all non-compact exceptional complex symmetric spaces . The results for G_2 are found in table I.

The following tables consist of the results of the algorithm described above applied to non-noticed even nilpotent orbits of $K_{\mathbb{C}}$ on $\mathfrak{p}_{\mathbb{C}}$. In the tables, when the nilpotent e is semiregular in the minimal regular subalgebra $\mathfrak{s}$ then e is given in terms of parameters u and v which can be determined by solving the system $x = [e, f]$. For more information on semiregular nilpotents in $\mathfrak{p}_{\mathbb{C}}$ the reader may consult Djoković [**D3**]. The orbit numbers and labels are found in [**D1**],[**D2**]. For each non-noticed even orbit of $\mathfrak{g}$ we give the Djoković's label, the normal triple (x, e, f), the maximal torus $\mathfrak{t}_{\mathbb{C}}^1$ and the type of $\mathfrak{k}_{\mathbb{C}}^{(x,e,f)}$.

Remark: When $\mathfrak{g}$ is of type EI or EIV then the nilpotent e is sometimes given in terms of the Cartan involution θ and a specific Chevalley basis is used. More details are found in [**No2**].

GI.

Let $\Delta = \{\alpha_1, \alpha_2\}$ be the Bourbaki simple roots of $\mathfrak{g}_{\mathbb{C}}$ then $\Delta_k = \{\beta_1, \beta_2\}$, where $\beta_1 = \alpha_1$ and $\beta_2 = 3\alpha_1 + 2\alpha_2$, is a set of simple roots for $\mathfrak{k}_{\mathbb{C}} = \mathfrak{sl}_2(\mathbb{C}) \oplus \mathfrak{sl}_2(\mathbb{C})$.

Table I

1. 1 1 $\mathfrak{k}_{\mathbb{C}}^{(x,e,f)} \simeq T_1$

$x = H_{\alpha_1} + H_{\alpha_2}, \quad e = X_{3\alpha_1+\alpha_2}$

$\mathfrak{t}_{\mathbb{C}}^1 = \mathbb{C}(H_{\alpha_1} + 3H_{\alpha_2})$

---

2. 1 3 $\mathfrak{k}_{\mathbb{C}}^{(x,e,f)} \simeq T_1$

$x = 2H_{\alpha_1} + 3H_{\alpha_2}, \quad e = X_{2\alpha_1+\alpha_2}$

$\mathfrak{t}_{\mathbb{C}}^1 = \mathbb{C}H_{\alpha_2}$

FI.

Let $\Delta = \{\alpha_1, \alpha_2, \alpha_3, \alpha_4\}$ be the Bourbaki simple roots of $\mathfrak{g}_{\mathbb{C}}$ then $\Delta_k = \{\beta_1, \ldots, \beta_4\}$, where $\beta_1 = \alpha_4$, $\beta_2 = \alpha_3$, $\beta_3 = \alpha_2$ and $\beta_4 = 2\alpha_1 + 3\alpha_2 + 4\alpha_3 + 2\alpha_4$, is a set of simple roots for $\mathfrak{k}_{\mathbb{C}} = \mathfrak{sp}_3(\mathbb{C}) \oplus \mathfrak{sl}_2(\mathbb{C})$.

Table II

---

6. 000 4 $\mathfrak{k}_{\mathbb{C}}^{(x,e,f)} \simeq A_2$

$x = 4H_{\alpha_1} + 6H_{\alpha_2} + 4H_{\alpha_3} + 2H_{\alpha_4}$, $e = \sqrt{2}(X_{\alpha_1} + X_{\alpha_1+3\alpha_2+4\alpha_3+2\alpha_4})$

$\mathfrak{t}_{\mathbb{C}}^1 = \mathbb{C}H_{\alpha_3} \oplus \mathbb{C}H_{\alpha_4}$

---

7. 200 0 $\mathfrak{k}_{\mathbb{C}}^{(x,e,f)} \simeq A_1 + T_1$

$x = 2H_{\alpha_2} + 2H_{\alpha_3} + 2H_{\alpha_4}$, $e = \sqrt{2}(X_{-\alpha_1} + X_{\alpha_1+\alpha_2+2\alpha_3+2\alpha_4})$

$\mathfrak{t}_{\mathbb{C}}^1 = \mathbb{C}H_{\alpha_3} \oplus \mathbb{C}(2H_{\alpha_1} + 4H_{\alpha_2} + H_{\alpha_4})$

---

8. 002 2 $\mathfrak{k}_{\mathbb{C}}^{(x,e,f)} \simeq A_1$

$x = 2H_{\alpha_1} + 6H_{\alpha_2} + 4H_{\alpha_3} + 2H_{\alpha_4}$

$e = X_{\alpha_1+\alpha_2+2\alpha_3} + X_{-\alpha_1} + X_{\alpha_1+2\alpha_2+3\alpha_3+2\alpha_4}$

$\mathfrak{t}_{\mathbb{C}}^1 = \mathbb{C}H_{\alpha_3}$

---

9. 020 0 $\mathfrak{k}_{\mathbb{C}}^{(x,e,f)} \simeq 2A_1$

$x = 4H_{\alpha_2} + 4H_{\alpha_3} + 2H_{\alpha_4}$

$e = \sqrt{2}(X_{\alpha_1+2\alpha_2+3\alpha_3+\alpha_4} + X_{-\alpha_1-\alpha_2-\alpha_3})$

$\mathfrak{t}_{\mathbb{C}}^1 = \mathbb{C}H_{\alpha_2} \oplus \mathbb{C}(H_{\alpha_1} + H_{\alpha_3} + H_{\alpha_4})$

---

19. 004 8 $\mathfrak{k}_{\mathbb{C}}^{(x,e,f)} \simeq A_1$

$x = 8H_{\alpha_1} + 18H_{\alpha_2} + 12H_{\alpha_3} + 6H_{\alpha_4}$

$e = \sqrt{6}(X_{\alpha_1+\alpha_2+2\alpha_3+2\alpha_4} + X_{\alpha_1+\alpha_2+\alpha_3}) + \sqrt{10}X_{-\alpha_1}$

$\mathfrak{t}^1_{\mathbb{C}} = \mathbb{C}H_{\alpha_3}$

---

20. 204 4 $\mathfrak{k}^{(x,e,f)}_{\mathbb{C}} \simeq T_1$

$x = 4H_{\alpha_1} + 14H_{\alpha_2} 10H_{\alpha_3} + 6H_{\alpha_4}$

$e = \sqrt{6}(X_{\alpha_1+\alpha_2+2\alpha_3+2\alpha_4} + X_{-\alpha_1-\alpha_2-\alpha_3}) + \sqrt{10}X_{\alpha_1+2\alpha_2+2\alpha_3}$

$\mathfrak{t}^1_{\mathbb{C}} = \mathbb{C}H_{\alpha_3}$

FII.

Let $\Delta = \{\alpha_1, \alpha_2, \alpha_3, \alpha_4\}$ be the Bourbaki system of simple roots of $\mathfrak{g}_{\mathbb{C}}$. Then $\Delta_k = \{\beta_1, \beta_2, \beta_3, \beta_4 : \beta_1 = \alpha_2 + 2\alpha_3 + 2\alpha_4, \beta_2 = \alpha_1, \beta_3 = \alpha_2, \beta_4 = \alpha_3, \}$ is a system of simple roots for $\mathfrak{k}_{\mathbb{C}} = \mathfrak{so}_4(\mathbb{C})$.

Table III

---

2. 4000 $\mathfrak{k}^{(x,e,f)}_{\mathbb{C}} \simeq G_2$

$x = 4H_{\alpha_1} + 8H_{\alpha_2} + 6H_{\alpha_3} + 4H_{\alpha_4}$, $e = \sqrt{2}(X_{\alpha_1+2\alpha_2+3\alpha_3+\alpha_4} + X_{\alpha_4})$

$\mathfrak{t}^1_{\mathbb{C}} = \mathbb{C}H_{\alpha_1} \oplus \mathbb{C}H_{\alpha_2}$

EI.

Let $\Delta = \{\alpha_1, \alpha_2, \ldots, \alpha_6\}$ the Bourbaki simple roots of $\mathfrak{g}_{\mathbb{C}}$ then $\Delta_k = \{\beta_0, \beta_4, \beta_3, \beta_2\}$, where $\beta_1 = \alpha_2$, $\beta_2 = \alpha_4$, $\beta_3 = \frac{\alpha_3+\alpha_5}{2}$, $\beta_4 = \frac{\alpha_1+\alpha_6}{2}$ and $\beta_0 = -(\beta_1 + 2\beta_2 + 3\beta_3 + 2\beta_4)$ is a set of simple roots for $\mathfrak{k}_{\mathbb{C}} = C_4$.

Table IV

---

4. 0002 $\mathfrak{k}^{(x,e,f)}_{\mathbb{C}} \simeq 2A_1$

$x = -2H_{\alpha_2}$

$e = (X_{-\alpha_2-\alpha_3-\alpha_4} - \theta X_{-\alpha_2-\alpha_3-\alpha_4}) + (X_{\alpha_4+\alpha_5} - \theta X_{\alpha_4+\alpha_5})$

$\mathfrak{t}^1_{\mathbb{C}} = \mathbb{C}(H_{\alpha_1} + H_{\alpha_2} + 2H_{\alpha_4} + H_{\alpha_6}) \oplus \mathbb{C}(H_{\alpha_3} + H_{\alpha_5})$

---

5. 2000 $\mathfrak{k}^{(x,e,f)}_{\mathbb{C}} \simeq 2A_1 + T2$

$x = -2H_{\alpha_1} - 4H_{\alpha_2} - 4H_{\alpha_3} - 6H_{\alpha_4} - 4H_{\alpha_5} - 2H_{\alpha_6}$

$e = \sqrt{2}(X_{-\alpha_1-\alpha_2-2\alpha_3-3\alpha_4-2\alpha_5-\alpha_6} + X_{-\alpha_2})$

$\mathfrak{t}^1_{\mathbb{C}} = \mathbb{C}H_{\alpha_1} \oplus \mathbb{C}H_{\alpha_3} \oplus \mathbb{C}H_{\alpha_5} \oplus \mathbb{C}H_{\alpha_6}$

---

6. 0200 $\mathfrak{k}_{\mathbb{C}}^{(x,e,f)} \simeq 2A_1$

$x = -4H_{\alpha_2} - 2H_{\alpha_3} - 4H_{\alpha_4} - 2H_{\alpha_5}$

$e = \sqrt{2}(X_{-\alpha_1-\alpha_2-\alpha_3-2\alpha_4-\alpha_5} - \theta X_{-\alpha_1-\alpha_2-\alpha_3-2\alpha_4-\alpha_5} + X_{\alpha_1+\alpha_3+\alpha_4} - \theta X_{\alpha_1+\alpha_3+\alpha_4})$

$\mathfrak{t}^1_{\mathbb{C}} = \mathbb{C}(H_{\alpha_1} + H_{\alpha_2} + H_{\alpha_6}) \oplus \mathbb{C}(H_{\alpha_3} + H_{\alpha_4} + H_{\alpha_5})$

---

9. 0202 $\mathfrak{k}_{\mathbb{C}}^{(x,e,f)} \simeq T_1$

$x = -6H_{\alpha_2} - 2H_{\alpha_3} - 4H_{\alpha_4} - 2H_{\alpha_5}$

$e = \sqrt{3}(X_{-\alpha_1-\alpha_2-\alpha_3-2\alpha_4-\alpha_5} - \theta X_{-\alpha_1-\alpha_2-\alpha_3-2\alpha_4-\alpha_5} + X_{\alpha_1} - \theta X_{\alpha_1})$
$\quad +2(X_{\alpha_3+\alpha_4} - \theta X_{\alpha_3+\alpha_4})$

$\mathfrak{t}^1_{\mathbb{C}} = \mathbb{C}(H_{\alpha_1} + H_{\alpha_2} + 2H_{\alpha_3} + 2H_{\alpha_4} + 2H_{\alpha_5} + H_{\alpha_6})$

---

12. 2002 $\mathfrak{k}_{\mathbb{C}}^{(x,e,f)} \simeq T_1$

$x = -2H_{\alpha_1} - 6H_{\alpha_2} - 4H_{\alpha_3} - 6H_{\alpha_4} - 4H_{\alpha_5} - 2H_{\alpha_6}$

$e = X_{-\alpha_1-\alpha_2-2\alpha_3-2\alpha_4-\alpha_5-\alpha_6} - \theta X_{-\alpha_1-\alpha_2-2\alpha_3-2\alpha_4-\alpha_5-\alpha_6} + 2X_{-\alpha_2-\alpha_3-2\alpha_4-\alpha_5}$
$\quad +\sqrt{3}(X_{\alpha_3+\alpha_4} - \theta X_{\alpha_3+\alpha_4})$

$\mathfrak{t}^1_{\mathbb{C}} = \mathbb{C}(H_{\alpha_3} + H_{\alpha_5})$

---

13. 2004 $\mathfrak{k}_{\mathbb{C}}^{(x,e,f)} \simeq A_1$

$x = -2H_{\alpha_1} - 8H_{\alpha_2} - 4H_{\alpha_3} - 6H_{\alpha_4} - 4H_{\alpha_5} - 2H_{\alpha_6}$

$e = \sqrt{6}X_{-\alpha_1-\alpha_2-\alpha_3-2\alpha_4-\alpha_5-\alpha_6} + \sqrt{10}X_{\alpha_1+\alpha_2+2\alpha_3+3\alpha_4+2\alpha_5+\alpha_6}$
$\quad +\sqrt{6}(X_{-\alpha_1-\alpha_2-2\alpha_3-2\alpha_4-\alpha_5} - \theta X_{-\alpha_1-\alpha_2-2\alpha_3-2\alpha_4-\alpha_5})$

$\mathfrak{t}^1_{\mathbb{C}} = \mathbb{C}(H_{\alpha_1} + H_{\alpha_3} + H_{\alpha_5} + H_{\alpha_6})$

EII.

Let $\Delta = \{\alpha_1, \alpha_2, \ldots, \alpha_6\}$ be the Bourbaki simple roots of $\mathfrak{g}_{\mathbb{C}}$ then $\Delta_k = \{\beta_1, \ldots, \beta_6\}$, where $\beta_1 = \alpha_1$, $\beta_2 = \alpha_3$, $\beta_3 = \alpha_4$, $\beta_4 = \alpha_5$, $\beta_5 = \alpha_6$ and $\beta_6 = \alpha_1 + 2\alpha_2 + 2\alpha_3 + 3\alpha_4 + 2\alpha_5 + \alpha_6$, is a set of simple roots for $\mathfrak{k}_{\mathbb{C}} = \mathfrak{sl}_6(\mathbb{C}) \oplus \mathfrak{sl}_2(\mathbb{C})$.

Table V

---

6. 00000 4 $\mathfrak{k}_{\mathbb{C}}^{(x,e,f)} \simeq 2A_2$

$x = 2H_{\alpha_1} + 4H_{\alpha_2} + 4H_{\alpha_3} + 6H_{\alpha_4} + 4H_{\alpha_5} + 2H_{\alpha_6}$

$e = \sqrt{2}(X_{\alpha_1+\alpha_2+\alpha_3+2\alpha_4+\alpha_5} + X_{\alpha_2+\alpha_3+\alpha_4+\alpha_5+\alpha_6})$

$\mathfrak{t}^1_{\mathbb{C}} = \mathbb{C}H_{\alpha_5} \oplus \mathbb{C}(H_{\alpha_1} + H_{\alpha_3}) \oplus \mathbb{C}(H_{\alpha_4} + H_{\alpha_6}) \oplus \mathbb{C}(H_{\alpha_3} + H_{\alpha_4})$

---

7. 20002 0 $\mathfrak{k}^{(x,e,f)}_{\mathbb{C}} \simeq 2A_1 + T_2$

$x = 2H_{\alpha_1} + 2H_{\alpha_3} + 2H_{\alpha_4} + 2H_{\alpha_5} + 2H_{\alpha_6}$

$e = \sqrt{2}(X_{\alpha_1+\alpha_2+\alpha_3+\alpha_4+\alpha_5+\alpha_6} + X_{-\alpha_2})$

$\mathfrak{t}^1_{\mathbb{C}} = \mathbb{C}H_{\alpha_3} \oplus \mathbb{C}H_{\alpha_5} \oplus \mathbb{C}(H_{\alpha_1} + H_{\alpha_2} + 2H_{\alpha_4}) \oplus \mathbb{C}(H_{\alpha_1} - H_{\alpha_6})$

---

8. 00200 2 $\mathfrak{k}^{(x,e,f)}_{\mathbb{C}} \simeq A_2$

$x = 2H_{\alpha_1} + 2H_{\alpha_2} + 4H_{\alpha_3} + 6H_{\alpha_4} + 4H_{\alpha_5} + 2H_{\alpha_6}$

$e = X_{\alpha_2+\alpha_3+2\alpha_4+2\alpha_5+\alpha_6} + X_{\alpha_1+\alpha_2+\alpha_3+2\alpha_4+\alpha_5+\alpha_6} + X_{-\alpha_2} + X_{\alpha_1+\alpha_2+2\alpha_3+2\alpha_4+\alpha_5}$

$\mathfrak{t}^1_{\mathbb{C}} = \mathbb{C}(H_{\alpha_1} + H_{\alpha_5}) \oplus \mathbb{C}(H_{\alpha_3} + H_{\alpha_6})$

---

11. 02020 0 $\mathfrak{k}^{(x,e,f)}_{\mathbb{C}} \simeq 2A_1$

$x = 2H_{\alpha_1} + 4H_{\alpha_3} + 4H_{\alpha_4} + 4H_{\alpha_5} + 2H_{\alpha_6}$

$e = \sqrt{2}(X_{\alpha_1+\alpha_2+2\alpha_3+2\alpha_4+\alpha_5} + X_{-\alpha_2-\alpha_3-\alpha_4})$
$\quad + \sqrt{2}(X_{\alpha_2+\alpha_3+2\alpha_4+2\alpha_5+\alpha_6} + X_{-\alpha_2-\alpha_4-\alpha_5})$

$\mathfrak{t}^1_{\mathbb{C}} = \mathbb{C}H_{\alpha_4} \oplus \mathbb{C}(H_{\alpha_1} + H_{\alpha_2} + H_{\alpha_3} + H_{\alpha_5} + H_{\alpha_6})$

---

20. 00400 0 $\mathfrak{k}^{(x,e,f)}_{\mathbb{C}} \simeq T_2$

$x = 2H_{\alpha_1} + 4H_{\alpha_3} + 6H_{\alpha_4} + 4H_{\alpha_5} + 2H_{\alpha_6}$

$e = \sqrt{4-u}X_{\alpha_2+\alpha_3+2\alpha_4+\alpha_5} + \sqrt{u-2}X_{\alpha_1+\alpha_2+\alpha_3+2\alpha_4+\alpha_5+\alpha_6} + \sqrt{6-u}X_{-\alpha_2-\alpha_3-\alpha_4-\alpha_5}$
$\quad + 2X_{\alpha_1+\alpha_2+2\alpha_3+2\alpha_4+2\alpha_5+\alpha_6} + \sqrt{e}X_{-\alpha_1-\alpha_2-\alpha_3-\alpha_4-\alpha_5-\alpha_6}$

$\mathfrak{t}^1_{\mathbb{C}} = \mathbb{C}(H_{\alpha_1} - H_{\alpha_6}) \oplus \mathbb{C}(H_{\alpha_3} - H_{\alpha_5})$

---

21. 02020 4 $\mathfrak{k}^{(x,e,f)}_{\mathbb{C}} \simeq T_2$

$x = 4H_{\alpha_1} + 4H_{\alpha_2} + 8H_{\alpha_3} + 10H_{\alpha_4} + 8H_{\alpha_5} + 4H_{\alpha_6}$

$e = \sqrt{4-u}X_{\alpha_2+\alpha_3+\alpha_4+\alpha_5} + \sqrt{u-2}X_{-\alpha_2-\alpha_4} + \sqrt{6-u}X_{-\alpha_2}$
$\quad + 2X_{\alpha_1+\alpha_2+\alpha_3+\alpha_4+\alpha_5+\alpha_6} + \sqrt{e}X_{\alpha_2+\alpha_3+2\alpha_4+\alpha_5}$

$\mathfrak{t}^1_{\mathbb{C}} = \mathbb{C}(H_{\alpha_1} - H_{\alpha_6}) \oplus \mathbb{C}(H_{\alpha_3} - H_{\alpha_5})$

---

22. 20202 2 $\mathfrak{k}^{(x,e,f)}_{\mathbb{C}} \simeq T_1$

$x = 4H_{\alpha_1} + 2H_{\alpha_2} + 6H_{\alpha_3} + 8H_{\alpha_4} + 6H_{\alpha_5} + 4H_{\alpha_6}$

$e = X_{\alpha_1+\alpha_2+\alpha_3+\alpha_4+\alpha_5+\alpha_6} + X_{-\alpha_2-\alpha_3-\alpha_4-\alpha_5} + 2X_{-\alpha_2-\alpha_4}$
$\quad + \sqrt{3}(X_{\alpha_1+\alpha_2+\alpha_3+2\alpha_4+\alpha_5} + X_{\alpha_2+\alpha_3+2\alpha_4+\alpha_5+\alpha_6})$

$\mathfrak{t}^1_{\mathbb{C}} = \mathbb{C}(H_{\alpha_1} + 2H_{\alpha_3} - 2H_{\alpha_5} - H_{\alpha_6})$

---

23. 00400 8 $\mathfrak{k}_{\mathbb{C}}^{(x,e,f)} \simeq A_2$

$$x = 6H_{\alpha_1} + 8H_{\alpha_2} + 12H_{\alpha_3} + 18H_{\alpha_4} + 12H_{\alpha_5} + 6H_{\alpha_6}$$

$$e = \sqrt{6}(X_{\alpha_1+\alpha_2+\alpha_3+\alpha_4} + X_{\alpha_2+\alpha_3+\alpha_4+\alpha_5} + X_{\alpha_2+\alpha_4+\alpha_5+\alpha_6}) + \sqrt{10}X_{-\alpha_2}$$

$$\mathfrak{t}_{\mathbb{C}}^1 = \mathbb{C}(H_{\alpha_1} + H_{\alpha_5}) \oplus \mathbb{C}(H_{\alpha_3} + H_{\alpha_6})$$

---

24. 20402 4 $\mathfrak{k}_{\mathbb{C}}^{(x,e,f)} \simeq A_1 + T_1$

$$x = 6H_{\alpha_1} + 4H_{\alpha_2} + 10H_{\alpha_3} + 14H_{\alpha_4} + 10H_{\alpha_5} + 6H_{\alpha_6}$$

$$e = \sqrt{10}X_{+\alpha_2+\alpha_3+2\alpha_4+\alpha_5} + \sqrt{6}(X_{\alpha_1+\alpha_2+\alpha_3+\alpha_4+\alpha_5+\alpha_6} + X_{-\alpha_2-\alpha_4} + X_{-\alpha_2-\alpha_3-\alpha_4-\alpha_5})$$

$$\mathfrak{t}_{\mathbb{C}}^1 = \mathbb{C}(H_{\alpha_1} - H_{\alpha_6}) \oplus \mathbb{C}(H_{\alpha_3} - H_{\alpha_5})$$

---

25. 40004 4 $\mathfrak{k}_{\mathbb{C}}^{(x,e,f)} \simeq A_1 + T_1$

$$x = 6H_{\alpha_1} + 4H_{\alpha_2} + 8H_{\alpha_3} + 10H_{\alpha_4} + 8H_{\alpha_5} + 6H_{\alpha_6}$$

$$e = \sqrt{6}(X_{\alpha_1+\alpha_2+\alpha_3+\alpha_4} + X_{\alpha_2+\alpha_3+2\alpha_4+2\alpha_5+\alpha_6}) \\ +2(X_{-\alpha_2-\alpha_3-2\alpha_4-\alpha_5} + X_{-\alpha_2})$$

$$\mathfrak{t}_{\mathbb{C}}^1 = \mathbb{C}H_{\alpha_3} \oplus \mathbb{C}(H_{\alpha_1} + H_{\alpha_5} - H_{\alpha_6})$$

---

26. 22022 0 $\mathfrak{k}_{\mathbb{C}}^{(x,e,f)} \simeq T_2$

$$x = 4H_{\alpha_1} + 6H_{\alpha_3} + 6H_{\alpha_4} + 6H_{\alpha_5} + 4H_{\alpha_6}$$

$$e = 2(X_{\alpha_1+\alpha_2+\alpha_3+2\alpha_4+\alpha_5+\alpha_6} + X_{-\alpha_2-\alpha_3-2\alpha_4-\alpha_5} \\ +\sqrt{6}(X_{\alpha_2+\alpha_3+2\alpha_4+2\alpha_5+\alpha_6} + X_{-\alpha_2-\alpha_4-\alpha_5-\alpha_6})$$

$$\mathfrak{t}_{\mathbb{C}}^1 = \mathbb{C}(H_{\alpha_1} + H_{\alpha_2} + H_{\alpha_3} + H_{\alpha_4} + H_{\alpha_5}) \oplus \mathbb{C}(H_{\alpha_2} + 2H_{\alpha_3} + H_{\alpha_4} + H_{\alpha_6})$$

---

34. 22422 4 $\mathfrak{k}_{\mathbb{C}}^{(x,e,f)} \simeq T_1$

$$x = 8H_{\alpha_1} + 4H_{\alpha_2} + 14H_{\alpha_3} + 18H_{\alpha_4} + 14H_{\alpha_5} + 8H_{\alpha_6}$$

$$e = \sqrt{10}X_{\alpha_1+\alpha_2+\alpha_3+\alpha_4+\alpha_5+\alpha_6} + 2X_{-\alpha_2-\alpha_3-\alpha_4-\alpha_5} \\ +\sqrt{6}(X_{\alpha_2+\alpha_3+2\alpha_4+\alpha_5} + X_{-\alpha_1-\alpha_2-\alpha_3-\alpha_4} + X_{-\alpha_2-\alpha_4-\alpha_5-\alpha_6})$$

$$\mathfrak{t}_{\mathbb{C}}^1 = \mathbb{C}(H_{\alpha_1} - H_{\alpha_3} + H_{\alpha_5} - H_{\alpha_6})$$

---

35. 40404 8 $\mathfrak{k}_{\mathbb{C}}^{(x,e,f)} \simeq T_1$

$$x = 10H_{\alpha_1} + 8H_{\alpha_2} + 16H_{\alpha_3} + 22H_{\alpha_4} + 16H_{\alpha_5} + 10H_{\alpha_6}$$

$$e = \sqrt{10}X_{\alpha_1+\alpha_2+\alpha_3+\alpha_4} + \sqrt{8}X_{-\alpha_2-\alpha_3-\alpha_4-\alpha_5} + \sqrt{10}X_{\alpha_2+\alpha_4+\alpha_5+\alpha_6} \\ +\sqrt{14}X_{\alpha_2+\alpha_3+2\alpha_4+\alpha_5} + \sqrt{18}X_{-\alpha_2-\alpha_4}$$

$$\mathfrak{t}_{\mathbb{C}}^1 = \mathbb{C}(H_{\alpha_1} - H_{\alpha_3} + H_{\alpha_5} - H_{\alpha_6})$$

EIII.

Let $\Delta = \{\alpha_1, \alpha_2, \ldots, \alpha_6\}$ be the Bourbaki simple roots of $\mathfrak{g}_{\mathbb{C}}$ then $\Delta_k = \{\beta_1, \ldots, \beta_6\}$, where $\beta_1 = \alpha_1$, $\beta_2 = \alpha_3$, $\beta_3 = \alpha_4$, $\beta_4 = \alpha_2$, $\beta_5 = \alpha_5$ and $\beta_6 = -\alpha_1 - 2\alpha_2 - 2\alpha_3 - 3\alpha_4 - 2\alpha_5 - \alpha_6$, is a set of simple roots for $\mathfrak{k}_{\mathbb{C}} = \mathfrak{so}_{10}(\mathbb{C}) \oplus \mathbb{C}$.

Table VI

6. 02000 -2 $\mathfrak{k}_{\mathbb{C}}^{(x,e,f)} \simeq A_2 + A_1 + T_1$

$$x = 2H_{\alpha_1} + 2H_{\alpha_2} + 4H_{\alpha_3} + 4H_{\alpha_4} + 2H_{\alpha_5}$$
$$e = \sqrt{2}(X_{\alpha_1+\alpha_2+2\alpha_3+2\alpha_4+\alpha_5+\alpha_6} + X_{-\alpha_6})$$
$$\mathfrak{t}_{\mathbb{C}}^1 = \mathbb{C}H_{\alpha_1} \oplus \mathbb{C}H_{\alpha_2} \oplus \mathbb{C}(H_{\alpha_3} + 2H_{\alpha_5} + H_{\alpha_6}) \oplus \mathbb{C}H_{\alpha_4}$$

9. 40000 -2 $\mathfrak{k}_{\mathbb{C}}^{(x,e,f)} \simeq G_2$

$$x = 4H_{\alpha_1} + 2H_{\alpha_2} + 4H_{\alpha_3} + 4H_{\alpha_4} + 2H_{\alpha_5}$$
$$e = \sqrt{2}(X_{\alpha_1+\alpha_3+\alpha_4+\alpha_5+\alpha_6} + X_{\alpha_1+2\alpha_2+2\alpha_3+3\alpha_4+2\alpha_5+\alpha_6}) + \sqrt{2}(X_{-\alpha_2-\alpha_3-2\alpha_4-2\alpha_5-\alpha_6} + X_{-\alpha_6})$$
$$\mathfrak{t}_{\mathbb{C}}^1 = \mathbb{C}H_{\alpha_3} \oplus \mathbb{C}H_{\alpha_4}$$

12. 02022 -6 $\mathfrak{k}_{\mathbb{C}}^{(x,e,f)} \simeq A_1 + T_1$

$$x = 4H_{\alpha_1} + 6H_{\alpha_2} + 8H_{\alpha_3} + 10H_{\alpha_4} + 6H_{\alpha_5}$$
$$e = +\sqrt{6}(X_{\alpha_2+\alpha_3+2\alpha_4+2\alpha_5+\alpha_6} + X_{-\alpha_3-\alpha_4-\alpha_5-\alpha_6}) + 2(X_{\alpha_1+\alpha_2+2\alpha_3+2\alpha_4+\alpha_5+\alpha_6} + X_{-\alpha_2-\alpha_4-\alpha_5-\alpha_6})$$
$$\mathfrak{t}_{\mathbb{C}}^1 = \mathbb{C}H_{\alpha_4} \oplus \mathbb{C}(2H_{\alpha_1} + H_{\alpha_3} + 2H_{\alpha_5} + H_{\alpha_6})$$

EIV.

Let $\Delta = \{\alpha_1, \alpha_2, \ldots, \alpha_6\}$ be the Bourbaki simple roots of $\mathfrak{g}_{\mathbb{C}}$ then $\Delta_k = \{\beta_1, \ldots, \beta_4\}$, where $\beta_1 = \alpha_2$, $\beta_2 = \alpha_4$, $\beta_3 = \frac{\alpha_3+\alpha_5}{2}$ and $\beta_4 = \frac{\alpha_1+\alpha_6}{2}$, is a set of simple roots for $\mathfrak{k}_{\mathbb{C}} = F_4$.

Table VII

2. 0002 $\mathfrak{k}_{\mathbb{C}}^{(x,e,f)} \simeq G_2$

$$x = 4H_{\alpha_1} + 4H_{\alpha_2} + 6H_{\alpha_3} + 8H_{\alpha_4} + 6H_{\alpha_5} + 4H_{\alpha_6}$$
$$e = \sqrt{2}(X_{\alpha_1} - \theta X_{\alpha_1} + X_{\alpha_1+\alpha_2+2\alpha_3+2\alpha_4+\alpha_5} - \theta X_{\alpha_1+\alpha_2+2\alpha_3+2\alpha_4+\alpha_5})$$
$$\mathfrak{t}_{\mathbb{C}}^1 = \mathbb{C}H_{\alpha_2} \oplus \mathbb{C}H_{\alpha_4}$$

EV.

Let $\Delta = \{\alpha_1, \alpha_2, \dots, \alpha_7\}$ be the Bourbaki simple roots of $\mathfrak{g}_{\mathbb{C}}$ then $\Delta_k = \{\beta_1, \dots, \beta_7\}$, where $\beta_1 = \alpha_1$, $\beta_2 = \alpha_3$, $\beta_3 = \alpha_4$, $\beta_4 = \alpha_5$, $\beta_5 = \alpha_6$, $\beta_6 = \alpha_7$ and $\beta_7 = \alpha_1 + 2\alpha_2 + 2\alpha_3 + 3\alpha_4 + 2\alpha_5 + \alpha_6$, is a set of simple roots for $\mathfrak{k}_{\mathbb{C}} = \mathfrak{sl}_8(\mathbb{C})$.

Table VIII

---

3. 0200000 $\quad \mathfrak{k}_{\mathbb{C}}^{(x,e,f)} \simeq C_3 + A_1$

$$x = 2H_{\alpha_1} + H_{\alpha_2} + 4H_{\alpha_3} + 4H_{\alpha_4} + 3H_{\alpha_5} + 2H_{\alpha_6} + H_{\alpha_7}$$

$$e = X_{\alpha_1+\alpha_2+2\alpha_3+2\alpha_4+\alpha_5} + X_{\alpha_1+\alpha_2+2\alpha_3+2\alpha_4+2\alpha_5+2\alpha_6+\alpha_7} + X_{-\alpha_2}$$

$$\mathfrak{t}_{\mathbb{C}}^1 = \mathbb{C}H_{\alpha_1} \oplus \mathbb{C}H_{\alpha_5} \oplus \mathbb{C}H_{\alpha_7} \oplus \mathbb{C}(H_{\alpha_2} + H_{\alpha_3} + 2H_{\alpha_4} + H_{\alpha_6})$$

---

4. 000020 $\quad \mathfrak{k}_{\mathbb{C}}^{(x,e,f)} \simeq C_3 + A_1$

$$x = 2H_{\alpha_1} + 3H_{\alpha_2} + 4H_{\alpha_3} + 6H_{\alpha_4} + 5H_{\alpha_5} + 4H_{\alpha_6} + 3H_{\alpha_7}$$

$$e = X_{\alpha_2+\alpha_4+\alpha_5+\alpha_6+\alpha_7} + X_{\alpha_1+\alpha_2+2\alpha_3+2\alpha_4+\alpha_5+\alpha_6+\alpha_7} + X_{\alpha_1+\alpha_2+2\alpha_3+3\alpha_4+3\alpha_5+2\alpha_6+\alpha_7}$$

$$\mathfrak{t}_{\mathbb{C}}^1 = \mathbb{C}H_{\alpha_1} \oplus \mathbb{C}H_{\alpha_4} \oplus \mathbb{C}H_{\alpha_6} \oplus \mathbb{C}(H_{\alpha_2} + H_{\alpha_3} + H_{\alpha_5})$$

---

6. 2000002 $\quad \mathfrak{k}_{\mathbb{C}}^{(x,e,f)} \simeq 2A_2 + T_1$

$$x = 4H_{\alpha_1} + 4H_{\alpha_2} + 6H_{\alpha_3} + 8H_{\alpha_4} + 6H_{\alpha_5} + 4H_{\alpha_6} + 2H_{\alpha_7}$$

$$e = \sqrt{2}(X_{\alpha_1+\alpha_2+\alpha_3+2\alpha_4+\alpha_5+\alpha_6+\alpha_7} + X_{\alpha_1+\alpha_2+2\alpha_3+2\alpha_4+2\alpha_5+\alpha_6})$$

$$\mathfrak{t}_{\mathbb{C}}^1 = \mathbb{C}H_{\alpha_2} \oplus \mathbb{C}H_{\alpha_6} \oplus \mathbb{C}(H_{\alpha_3} + H_{\alpha_4}) \oplus \mathbb{C}(H_{\alpha_3} - H_{\alpha_5}) \oplus \mathbb{C}(H_{\alpha_5} + H_{\alpha_7})$$

---

7. 0002000 $\quad \mathfrak{k}_{\mathbb{C}}^{(x,e,f)} \simeq A_3$

$$x = 2H_{\alpha_1} + 2H_{\alpha_2} + 4H_{\alpha_3} + 6H_{\alpha_4} + 6H_{\alpha_5} + 4H_{\alpha_6} + 2H_{\alpha_7}$$

$$e = X_{\alpha_1+\alpha_2+2\alpha_3+2\alpha_4+2\alpha_5+2\alpha_6+\alpha_7} + X_{\alpha_1+\alpha_2+\alpha_3+2\alpha_4+2\alpha_5+\alpha_6+\alpha_7} + X_{\alpha_2+\alpha_3+2\alpha_4+2\alpha_5+\alpha_6} + X_{-\alpha_2}$$

$$\mathfrak{t}_{\mathbb{C}}^1 = \mathbb{C}(H_{\alpha_1} + H_{\alpha_2} + H_{\alpha_3} + 2H_{\alpha_4} + H_{\alpha_5} + H_{\alpha_6}) \oplus \mathbb{C}(H_{\alpha_1} - H_{\alpha_7}) \oplus \mathbb{C}(H_{\alpha_3} - H_{\alpha_6})$$

---

16. 4000000 $\quad \mathfrak{k}_{\mathbb{C}}^{(x,e,f)} \simeq G_2$

$$x = 4H_{\alpha_1} + H_{\alpha_2} + 4H_{\alpha_3} + 4H_{\alpha_4} + 3H_{\alpha_5} + 2H_{\alpha_6} + H_{\alpha_7}$$

$$e = \sqrt{2}(X_{\alpha_1+\alpha_2+\alpha_3+\alpha_4} + X_{\alpha_1+\alpha_2+2\alpha_3+3\alpha_4+3\alpha_5+2\alpha_6+\alpha_7}) + X_{-\alpha_2-\alpha_4-\alpha_5} + X_{-\alpha_2-\alpha_3-\alpha_4-\alpha_5-\alpha_6} + X_{-\alpha_2-\alpha_3-2\alpha_4-\alpha_5-\alpha_6-\alpha_7}$$

$$\mathfrak{t}_{\mathbb{C}}^1 = \mathbb{C}(H_{\alpha_3} - H_{\alpha_6}) \oplus \mathbb{C}(H_{\alpha_4} - H_{\alpha_7})$$

---

17. 0000004 $\quad \mathfrak{k}_{\mathbb{C}}^{(x,e,f)} \simeq G_2$

$$x = 4H_{\alpha_1} + 7H_{\alpha_2} + 8H_{\alpha_3} + 12H_{\alpha_4} + 9H_{\alpha_5} + 6H_{\alpha_6} + 3H_{\alpha_7}$$

$$e = \sqrt{2}(X_{\alpha_2+\alpha_4} + X_{\alpha_1+\alpha_2+2\alpha_3+2\alpha_4+2\alpha_5+\alpha_6}) + X_{\alpha_1+\alpha_2+2\alpha_3+2\alpha_4+\alpha_5+\alpha_6+\alpha_7}$$

$+X_{\alpha_1+\alpha_2+\alpha_3+2\alpha_4+2\alpha_5+\alpha_6+\alpha_7} + X_{\alpha_2+\alpha_3+2\alpha_4+2\alpha_5+2\alpha_6+\alpha_7}$

$\mathfrak{t}^1_{\mathbb{C}} = \mathbb{C}(H_{\alpha_1} + H_{\alpha_6}) \oplus \mathbb{C}(H_{\alpha_3} + 2H_{\alpha_4} + H_{\alpha_5})$

---

18. 2000200 $\quad \mathfrak{k}_{\mathbb{C}}^{(x,e,f)} \simeq 2A_1$

$x = 4H_{\alpha_1} + 3H_{\alpha_2} + 6H_{\alpha_3} + 8H_{\alpha_4} + 7H_{\alpha_5} + 6H_{\alpha_6} + 3H_{\alpha_7}$

$e = \sqrt{2}(X_{\alpha_1+\alpha_2+2\alpha_3+2\alpha_4+\alpha_5+\alpha_6} + X_{\alpha_1+\alpha_2+\alpha_3+2\alpha_4+2\alpha_5+\alpha_6+\alpha_7})$
$\quad +X_{\alpha_2+\alpha_3+2\alpha_4+2\alpha_5+2\alpha_6+\alpha_7} + X_{-\alpha_2} + X_{-\alpha_2-\alpha_3-2\alpha_4-\alpha_5}$

$\mathfrak{t}^1_{\mathbb{C}} = \mathbb{C}(H_{\alpha_3} + H_{\alpha_5}) \oplus \mathbb{C}(H_{\alpha_3} + H_{\alpha_7})$

---

19. 0020002 $\quad \mathfrak{k}_{\mathbb{C}}^{(x,e,f)} \simeq 2A_1$

$x = 4H_{\alpha_1} + 5H_{\alpha_2} + 8H_{\alpha_3} + 12H_{\alpha_4} + 9H_{\alpha_5} + 6H_{\alpha_6} + 3H_{\alpha_7}$

$e = \sqrt{2}(X_{\alpha_1+\alpha_2+\alpha_3+2\alpha_4+\alpha_5+\alpha_6} + X_{\alpha_2+\alpha_3+2\alpha_4+2\alpha_5+\alpha_6+\alpha_7})$
$\quad +X_{\alpha_1+\alpha_2+2\alpha_3+2\alpha_4+2\alpha_5+\alpha_6} + X_{\alpha_1+\alpha_2+2\alpha_3+2\alpha_4+\alpha_5+\alpha_6+\alpha_7} + X_{-\alpha_2}$

$\mathfrak{t}^1_{\mathbb{C}} = \mathbb{C}(H_{\alpha_1} - H_{\alpha_6}) \oplus \mathbb{C}(2H_{\alpha_1} + H_{\alpha_5} + H_{\alpha_7})$

---

21. 0200020 $\quad \mathfrak{k}_{\mathbb{C}}^{(x,e,f)} \simeq 2A_1 + T_1$

$x = 4H_{\alpha_1} + 4H_{\alpha_2} + 8H_{\alpha_3} + 10H_{\alpha_4} + 8H_{\alpha_5} + 6H_{\alpha_6} + 4H_{\alpha_7}$

$e = \sqrt{2}(X_{\alpha_2+\alpha_3+\alpha_4+\alpha_5+\alpha_6+\alpha_7} + X_{\alpha_1+\alpha_2+2\alpha_3+2\alpha_4+\alpha_5+\alpha_6}$
$\quad +X_{\alpha_1+\alpha_2+\alpha_3+2\alpha_4+2\alpha_5+\alpha_6+\alpha_7} + X_{-\alpha_2})$

$\mathfrak{t}^1_{\mathbb{C}} = \mathbb{C}(H_{\alpha_2} + H_{\alpha_3} + 2H_{\alpha_4} + H_{\alpha_5}) \oplus \mathbb{C}(H_{\alpha_2} + H_{\alpha_3} + 2H_{\alpha_4} - H_{\alpha_6})$
$\quad \oplus\mathbb{C}(2H_{\alpha_2} + H_{\alpha_3} + 4H_{\alpha_4} + H_{\alpha_7})$

---

22. 0202000 $\quad \mathfrak{k}_{\mathbb{C}}^{(x,e,f)} \simeq 2A_1 + T_1$

$x = 4H_{\alpha_1} + 3H_{\alpha_2} + 8H_{\alpha_3} + 10H_{\alpha_4} + 9H_{\alpha_5} + 6H_{\alpha_6} + 3H_{\alpha_7}$

$e = \sqrt{3}(X_{\alpha_2+\alpha_3+2\alpha_4+2\alpha_5+\alpha_6} + X_{-\alpha_2+-\alpha_3-\alpha_4})$
$\quad +2X_{\alpha_1+\alpha_2+2\alpha_3+2\alpha_4+\alpha_5+\alpha_6+\alpha_7} + X_{-\alpha_2-\alpha_4-\alpha_5-\alpha_6-\alpha_7}$

$\mathfrak{t}^1_{\mathbb{C}} = \mathbb{C}H_{\alpha_4} \oplus \mathbb{C}H_{\alpha_6} \oplus \mathbb{C}(H_{\alpha_1} + H_{\alpha_2} + H_{\alpha_3} + H_{\alpha_5})$

---

23. 0002020 $\quad \mathfrak{k}_{\mathbb{C}}^{(x,e,f)} \simeq 3A_1$

$x = 4H_{\alpha_1} + 5H_{\alpha_2} + 8H_{\alpha_3} + 12H_{\alpha_4} + 11H_{\alpha_5} + 8H_{\alpha_6} + 5H_{\alpha_7}$

$e = 2X_{\alpha_1+\alpha_2+\alpha_3+\alpha_4+\alpha_5+\alpha_6+\alpha_7} + X_{\alpha_2+\alpha_3+2\alpha_4+\alpha_5+\alpha_6+\alpha_7}$
$\quad +\sqrt{3}(X_{\alpha_1+\alpha_2+2\alpha_3+3\alpha_4+2\alpha_5+\alpha_6} + X_{-\alpha_1-\alpha_2-\alpha_3-\alpha_4})$

$\mathfrak{t}^1_{\mathbb{C}} = \mathbb{C}H_{\alpha_3} \oplus \mathbb{C}H_{\alpha_6} \oplus \mathbb{C}(H_{\alpha_2} + H_{\alpha_4} + H_{\alpha_5})$

---

26. 2002002 $\quad \mathfrak{k}_{\mathbb{C}}^{(x,e,f)} \simeq A_1 + T_1$

$x = 6H_{\alpha_1} + 6H_{\alpha_2} + 10H_{\alpha_3} + 14H_{\alpha_4} + 12H_{\alpha_5} + 8H_{\alpha_6} + 4H_{\alpha_7}$

$e = X_{\alpha_1+\alpha_2+\alpha_3+\alpha_4+\alpha_5} + X_{\alpha_1+\alpha_2+\alpha_3+2\alpha_4+\alpha_5+\alpha_6}$
$\quad +\sqrt{3}(X_{\alpha_2+\alpha_3+2\alpha_4+2\alpha_5+\alpha_6} + X_{-\alpha_2-\alpha_3-\alpha_4}) + 2X_{\alpha_1+\alpha_2+2\alpha_3+2\alpha_4+\alpha_5+\alpha_6+\alpha_7}$

$\mathfrak{t}^1_{\mathbb{C}} = \mathbb{C}(H_{\alpha_4} - H_{\alpha_6}) \oplus \mathbb{C}(H_{\alpha_2} + 2H_{\alpha_4} + H_{\alpha_5} + H_{\alpha_7})$

---

27. 0020200 $\mathfrak{k}_{\mathbb{C}}^{(x,e,f)} \simeq T_3$

$$x = 4H_{\alpha_1} + 4H_{\alpha_2} + 8H_{\alpha_3} + 12H_{\alpha_4} + 10H_{\alpha_5} + 8H_{\alpha_6} + 4H_{\alpha_7}$$

$$e = \sqrt{4-u}X_{\alpha_2+\alpha_3+2\alpha_4+\alpha_5+\alpha_6} + 2X_{\alpha_1+\alpha_2+\alpha_3+2\alpha_4+2\alpha_5+\alpha_6+\alpha_7} + \sqrt{2-u}X_{-\alpha_2-\alpha_4} + \sqrt{u}X_{\alpha_2+\alpha_3+2\alpha_4+2\alpha_5+\alpha_6} + \sqrt{2+u}X_{-\alpha_2-\alpha_4-\alpha_5}$$

$$\mathfrak{t}_{\mathbb{C}}^1 = \mathbb{C}(H_{\alpha_1}+H_{\alpha_2}+2H_{\alpha_3}+2H_{\alpha_4}+H_{\alpha_5}) \oplus \mathbb{C}(H_{\alpha_1}-H_{\alpha_7}) \oplus \mathbb{C}(H_{\alpha_3}+H_{\alpha_4}-H_{\alpha_6})$$

---

30. 2004002 $\mathfrak{k}_{\mathbb{C}}^{(x,e,f)} \simeq A_2 + T_1$

$$x = 8H_{\alpha_1} + 8H_{\alpha_2} + 14H_{\alpha_3} + 20H_{\alpha_4} + 18H_{\alpha_5} + 12H_{\alpha_6} + 6H_{\alpha_7}$$

$$e = \sqrt{6}(X_{\alpha_1+\alpha_2+\alpha_3+\alpha_4+\alpha_5} + X_{\alpha_1+\alpha_2+\alpha_3+2\alpha_4+\alpha_5+\alpha_6} + X_{\alpha_1+\alpha_2+2\alpha_3+2\alpha_4+\alpha_5+\alpha_6+\alpha_7}) + \sqrt{10}X_{-\alpha_1-\alpha_2-\alpha_3-\alpha_4}$$

$$\mathfrak{t}_{\mathbb{C}}^1 = \mathbb{C}(H_{\alpha_3} - H_{\alpha_7}) \oplus \mathbb{C}(H_{\alpha_4} - H_{\alpha_6}) \oplus \mathbb{C}(H_{\alpha_2} + H_{\alpha_3} + 2H_{\alpha_4} + H_{\alpha_5})$$

---

38. 2200022 $\mathfrak{k}_{\mathbb{C}}^{(x,e,f)} \simeq A_1 + T_2$

$$x = 8H_{\alpha_1} + 8H_{\alpha_2} + 14H_{\alpha_3} + 18H_{\alpha_4} + 14H_{\alpha_5} + 10H_{\alpha_6} + 6H_{\alpha_7}$$

$$e = 2(X_{\alpha_1+\alpha_2+\alpha_3+2\alpha_4+\alpha_5} + X_{\alpha_1+\alpha_2+\alpha_3+\alpha_4+\alpha_5+\alpha_6}) + \sqrt{6}(X_{\alpha_2+\alpha_3+\alpha_4+\alpha_5+\alpha_6+\alpha_7} + X_{-\alpha_2})$$

$$\mathfrak{t}_{\mathbb{C}}^1 = \mathbb{C}H_{\alpha_5} \oplus \mathbb{C}(H_{\alpha_2} + H_{\alpha_3} + 2H_{\alpha_4} + H_{\alpha_6}) \oplus \mathbb{C}(H_{\alpha_2} + 2H_{\alpha_3} + 2H_{\alpha_4} - H_{\alpha_7})$$

---

39. 0040000 $\mathfrak{k}_{\mathbb{C}}^{(x,e,f)} \simeq A_1$

$$x = 4H_{\alpha_1} + 3H_{\alpha_2} + 8H_{\alpha_3} + 12H_{\alpha_4} + 9H_{\alpha_5} + 6H_{\alpha_6} + 3H_{\alpha_7}$$

$$e = \sqrt{2}X_{+\alpha_2+\alpha_3+2\alpha_4+\alpha_5} + 2X_{\alpha_1+\alpha_2+\alpha_3+2\alpha_4+\alpha_5+\alpha_6} + X_{\alpha_1+\alpha_2+\alpha_3+2\alpha_4+2\alpha_5+\alpha_6+\alpha_7} + \sqrt{2}X_{\alpha_1+\alpha_2+2\alpha_3+2\alpha_4+2\alpha_5+2\alpha_6+\alpha_7} + \sqrt{3}(X_{-\alpha_2-\alpha_4-\alpha_5-\alpha_6} + X_{-\alpha_1-\alpha_2-\alpha_3-\alpha_4})$$

$$\mathfrak{t}_{\mathbb{C}}^1 = \mathbb{C}(H_{\alpha_1} + H_{\alpha_3} + H_{\alpha_5} - H_{\alpha_6} - 2H_{\alpha_7})$$

---

40. 0040000 $\mathfrak{k}_{\mathbb{C}}^{(x,e,f)} \simeq A_1$

$$x = 4H_{\alpha_1} + 5H_{\alpha_2} + 8H_{\alpha_3} + 12H_{\alpha_4} + 11H_{\alpha_5} + 10H_{\alpha_6} + 5H_{\alpha_7}$$

$$e = \sqrt{2}X_{\alpha_2+\alpha_4+\alpha_5+\alpha_6} + 2X_{\alpha_1+\alpha_2+\alpha_3+\alpha_4+\alpha_5+\alpha_6+\alpha_7} + X_{\alpha_2+\alpha_3+2\alpha_4+2\alpha_5+\alpha_6+\alpha_7} + \sqrt{3}X_{\alpha_1+\alpha_2+2\alpha_3+3\alpha_4+2\alpha_5+\alpha_6} + \sqrt{2}X_{-\alpha_2-\alpha_4} + \sqrt{3}X_{-\alpha_1-\alpha_2-\alpha_3-\alpha_4-\alpha_5}$$

$$\mathfrak{t}_{\mathbb{C}}^1 = \mathbb{C}(2H_{\alpha_2} + 3H_{\alpha_3} + 2H_{\alpha_4} + H_{\alpha_5} + H_{\alpha_6})$$

---

41. 2020020 $\mathfrak{k}_{\mathbb{C}}^{(x,e,f)} \simeq T_1$

$$x = 6H_{\alpha_1} + 5H_{\alpha_2} + 10H_{\alpha_3} + 14H_{\alpha_4} + 11H_{\alpha_5} + 8H_{\alpha_6} + 5H_{\alpha_7}$$

$$e = X_{\alpha_1+\alpha_2+\alpha_3+\alpha_4+\alpha_5+\alpha_6+\alpha_7} + \sqrt{3}X_{\alpha_1+\alpha_2+2\alpha_3+2\alpha_4+\alpha_5+\alpha_6} + \sqrt{2}X_{\alpha_1+\alpha_2+\alpha_3+2\alpha_4+2\alpha_5+\alpha_6} + 2X_{\alpha_2+\alpha_3+2\alpha_4+2\alpha_5+\alpha_6+\alpha_7} + \sqrt{3}X_{-\alpha_2-\alpha_3-\alpha_4-\alpha_5} + \sqrt{2}X_{-\alpha_2-\alpha_4-\alpha_5-\alpha_6}$$

$$\mathfrak{t}_{\mathbb{C}}^1 = \mathbb{C}(H_{\alpha_1} + 2H_{\alpha_2} + H_{\alpha_3} + 3H_{\alpha_4} - H_{\alpha_6})$$

---

42. 0200202 $\mathfrak{k}_{\mathbb{C}}^{(x,e,f)} \simeq T_1$

$$x = 6H_{\alpha_1} + 7H_{\alpha_2} + 12H_{\alpha_3} + 16H_{\alpha_4} + 13H_{\alpha_5} + 10H_{\alpha_6} + 5H_{\alpha_7}$$

$$e = X_{\alpha_1+\alpha_2+2\alpha_3+2\alpha_4+\alpha_5} + \sqrt{3}X_{\alpha_1+\alpha_2+\alpha_3+2\alpha_4+\alpha_5+\alpha_6} + \sqrt{2}X_{\alpha_1+\alpha_2+\alpha_3+\alpha_4+\alpha_5+\alpha_6+\alpha_7} + \sqrt{2} + X_{\alpha_2+\alpha_3+2\alpha_4+2\alpha_5+\alpha_6} + \sqrt{3}X_{\alpha_2+\alpha_3+2\alpha_4+\alpha_5+\alpha_6+\alpha_7} + 2X_{-\alpha_2-\alpha_4}$$

$$\mathfrak{t}^1_{\mathbb{C}} = \mathbb{C}(H_{\alpha_1} + 2H_{\alpha_4} + 2H_{\alpha_5} + H_{\alpha_7})$$

---

43. 0202020 $\mathfrak{k}^{(x,e,f)}_{\mathbb{C}} \simeq A_1$

$$x = 6H_{\alpha_1} + 6H_{\alpha_2} + 12H_{\alpha_3} + 16H_{\alpha_4} + 14H_{\alpha_5} + 10H_{\alpha_6} + 6H_{\alpha_7}$$

$$e = \sqrt{3}(X_{\alpha_2+\alpha_3+\alpha_4+\alpha_5+\alpha_6+\alpha_7} + X_{\alpha_1+\alpha_2+2\alpha_3+2\alpha_4+\alpha_5}) + 2X_{\alpha_2+\alpha_3+2\alpha_4+2\alpha_5+\alpha_6} + \sqrt{3}(X_{\alpha_1+\alpha_2+\alpha_3+2\alpha_4+\alpha_5+\alpha_6+\alpha_7} + X_{-\alpha_2-\alpha_4-\alpha_5}) + 2X_{-\alpha_2-\alpha_3-\alpha_4}$$

$$\mathfrak{t}^1_{\mathbb{C}} = \mathbb{C}(H_{\alpha_1} + H_{\alpha_2} + H_{\alpha_3} + H_{\alpha_4} + H_{\alpha_5} + H_{\alpha_6})$$

---

44. 0402020 $\mathfrak{k}^{(x,e,f)}_{\mathbb{C}} \simeq 2A_1$

$$x = 8H_{\alpha_1} + 7H_{\alpha_2} + 16H_{\alpha_3} + 20H_{\alpha_4} + 17H_{\alpha_5} + 12H_{\alpha_6} + 7H_{\alpha_7}$$

$$e = \sqrt{8}X_{\alpha_1+\alpha_2+\alpha_3+\alpha_4+\alpha_5+\alpha_6+\alpha_7} + \sqrt{5}X_{\alpha_2+\alpha_3+2\alpha_4+2\alpha_5+\alpha_6}) + \sqrt{8}X_{+\alpha_2+\alpha_3+2\alpha_4+\alpha_5+\alpha_6+\alpha_7} + \sqrt{5}X_{-\alpha_2-\alpha_3-\alpha_4} + 3X_{-\alpha_2-\alpha_4-\alpha_5-\alpha_6-\alpha_7}$$

$$\mathfrak{t}^1_{\mathbb{C}} = \mathbb{C}H_{\alpha_6} \oplus \mathbb{C}(H_{\alpha_1} + H_{\alpha_2} + H_{\alpha_3} + 2H_{\alpha_4} + H_{\alpha_5})$$

---

45. 0202040 $\mathfrak{k}^{(x,e,f)}_{\mathbb{C}} \simeq 2A_1$

$$x = 8H_{\alpha_1} + 9H_{\alpha_2} + 16H_{\alpha_3} + 22H_{\alpha_4} + 19H_{\alpha_5} + 14H_{\alpha_6} + 9H_{\alpha_7}$$

$$e = 3X_{\alpha_2+\alpha_4+\alpha_5+\alpha_6+\alpha_7} + \sqrt{8}X_{\alpha_1+\alpha_2+2\alpha_3+2\alpha_4+\alpha_5} + \sqrt{5}X_{\alpha_1+\alpha_2+\alpha_3+2\alpha_4+2\alpha_5+\alpha_6} + \sqrt{8}X_{-\alpha_2-\alpha_4-\alpha_5} + \sqrt{5}X_{-\alpha_1-\alpha_2-\alpha_3-\alpha_4}$$

$$\mathfrak{t}^1_{\mathbb{C}} = \mathbb{C}H_{\alpha_4} \oplus \mathbb{C}(H_{\alpha_2} + H_{\alpha_3} + H_{\alpha_5} + H_{\alpha_6})$$

---

54. 2020202 $\mathfrak{k}^{(x,e,f)}_{\mathbb{C}} \simeq T_1$

$$x = 8H_{\alpha_1} + 8H_{\alpha_2} + 14H_{\alpha_3} + 20H_{\alpha_4} + 16H_{\alpha_5} + 12H_{\alpha_6} + 6H_{\alpha_7}$$

$$e = 2(X_{\alpha_1+\alpha_2+\alpha_3+\alpha_4+\alpha_5+\alpha_6} + X_{\alpha_1+\alpha_2+2\alpha_3+2\alpha_4+\alpha_5}) + \sqrt{2}X_{\alpha_2+\alpha_3+2\alpha_4+2\alpha_5+\alpha_6} + \sqrt{6}(X_{\alpha_2+\alpha_3+2\alpha_4+\alpha_5+\alpha_6+\alpha_7} + X_{-\alpha_2-\alpha_3-\alpha_4}) + \sqrt{2}X_{-\alpha_2-\alpha_4-\alpha_5}$$

$$\mathfrak{t}^1_{\mathbb{C}} = \mathbb{C}(H_{\alpha_2} + 2H_{\alpha_3} + 3H_{\alpha_5} + 2H_{\alpha_6} + 3H_{\alpha_7})$$

---

55. 4004000 $\mathfrak{k}^{(x,e,f)}_{\mathbb{C}} \simeq A_1$

$$x = 8H_{\alpha_1} + 5H_{\alpha_2} + 12H_{\alpha_3} + 16H_{\alpha_4} + 15H_{\alpha_5} + 10H_{\alpha_6} + 5H_{\alpha_7}$$

$$e = \sqrt{6}X_{\alpha_1+\alpha_2+\alpha_3+\alpha_4+\alpha_5} + \sqrt{7}X_{\alpha_1+\alpha_2+\alpha_3+2\alpha_4+\alpha_5+\alpha_6} + \sqrt{7-u}X_{\alpha_1+\alpha_2+2\alpha_3+2\alpha_4+\alpha_5+\alpha_6} + \sqrt{u-2}X_{\alpha_2+\alpha_3+2\alpha_4+2\alpha_5+2\alpha_6+\alpha_7} + \sqrt{12-u}X_{-\alpha_1-\alpha_2-\alpha_3-\alpha_4} + X_{-\alpha_2-\alpha_3-2\alpha_4-\alpha_5} + \sqrt{u}X_{-\alpha_2-\alpha_4-\alpha_5-\alpha_6}$$

$$\mathfrak{t}^1_{\mathbb{C}} = \mathbb{C}(H_{\alpha_3} - H_{\alpha_7})$$

---

56. 0004004 $\mathfrak{k}^{(x,e,f)}_{\mathbb{C}} \simeq A_1$

$$x = 8H_{\alpha_1} + 11H_{\alpha_2} + 16H_{\alpha_3} + 24H_{\alpha_4} + 21H_{\alpha_5} + 14H_{\alpha_6} + 7H_{\alpha_7}$$

$$e = \sqrt{7-u}X_{\alpha_2+\alpha_4+\alpha_5+\alpha_6+\alpha_7} + \sqrt{6}X_{\alpha_1+\alpha_2+\alpha_3+\alpha_4+\alpha_5+\alpha_6} + X_{\alpha_2+\alpha_3+2\alpha_4+\alpha_5+\alpha_6} + \sqrt{7}X_{\alpha_1+\alpha_2+2\alpha_3+2\alpha_4+\alpha_5} + \sqrt{5-u}X_{-\alpha_2} + \sqrt{5+u}X_{-\alpha_1-\alpha_2-\alpha_3-\alpha_4} + \sqrt{u}X_{\alpha_1+\alpha_2+\alpha_3+2\alpha_4+\alpha_5+\alpha_6+\alpha_7}$$

$$\mathfrak{t}^1_{\mathbb{C}} = \mathbb{C}(H_{\alpha_3} + H_{\alpha_6} + H_{\alpha_7})$$

---

57. 2022020 $\mathfrak{k}_{\mathbb{C}}^{(x,e,f)} \simeq T_1$

$$x = 8H_{\alpha_1} + 7H_{\alpha_2} + 14H_{\alpha_3} + 20H_{\alpha_4} + 17H_{\alpha_5} + 12H_{\alpha_6} + 7H_{\alpha_7}$$

$$e = \sqrt{12-u}X_{\alpha_1+\alpha_2+\alpha_3+2\alpha_4+\alpha_5+\alpha_6} + \sqrt{6}X_{\alpha_1+\alpha_2+\alpha_3+\alpha_4+\alpha_5+\alpha_6+\alpha_7} + X_{\alpha_2+\alpha_3+2\alpha_4+\alpha_5+\alpha_6+\alpha_7} + \sqrt{7}X_{-\alpha_2-\alpha_4-\alpha_5-\alpha_6} + \sqrt{u-5}X_{-\alpha_2-\alpha_3-\alpha_4-\alpha_5} + \sqrt{10-u}X_{-\alpha_1-\alpha_2-\alpha_3-\alpha_4} + \sqrt{u}X_{\alpha_2+\alpha_3+2\alpha_4+2\alpha_5+\alpha_6}$$

$$\mathfrak{t}^1_{\mathbb{C}} = \mathbb{C}(H_{\alpha_3} + H_{\alpha_6})$$

---

58. 0202202 $\mathfrak{k}_{\mathbb{C}}^{(x,e,f)} \simeq T_1$

$$x = 8H_{\alpha_1} + 9H_{\alpha_2} + 16H_{\alpha_3} + 22H_{\alpha_4} + 19H_{\alpha_5} + 14H_{\alpha_6} + 7H_{\alpha_7}$$

$$e = \sqrt{7-u}X_{\alpha_2+\alpha_3+\alpha_4+\alpha_5+\alpha_6} + \sqrt{6}X_{\alpha_1+\alpha_2+2\alpha_3+2\alpha_4+\alpha_5} + \sqrt{5+u}X_{\alpha_2+\alpha_3+2\alpha_4+\alpha_5+\alpha_6+\alpha_7} + X_{-\alpha_2-\alpha_4-\alpha_5} + \sqrt{10}X_{-\alpha_2-\alpha_3-\alpha_4} + \sqrt{u}X_{\alpha_1+\alpha_2+\alpha_3+\alpha_4+\alpha_5+\alpha_6} + \sqrt{2-u}X_{\alpha_1+\alpha_2+\alpha_3+2\alpha_4+\alpha_5+\alpha_6+\alpha_7}$$

$$\mathfrak{t}^1_{\mathbb{C}} = \mathbb{C}(H_{\alpha_4} - H_{\alpha_7})$$

---

62. 2202022 $\mathfrak{k}_{\mathbb{C}}^{(x,e,f)} \simeq T_1$

$$x = 10H_{\alpha_1} + 10H_{\alpha_2} + 18H_{\alpha_3} + 24H_{\alpha_4} + 20H_{\alpha_5} + 14H_{\alpha_6} + 8H_{\alpha_7}$$

$$e = X_{\alpha_1+\alpha_2+\alpha_3+\alpha_4+\alpha_5} + \sqrt{8}X_{\alpha_2+\alpha_3+\alpha_4+\alpha_5+\alpha_6+\alpha_7} + 3X_{\alpha_1+\alpha_2+\alpha_3+2\alpha_4+\alpha_5+\alpha_6} + \sqrt{5}(X_{\alpha_2+\alpha_3+2\alpha_4+2\alpha_5+\alpha_6} + X_{-\alpha_2-\alpha_3-\alpha_4}) + \sqrt{8}X_{-\alpha_2-\alpha_4-\alpha_5-\alpha_6}$$

$$\mathfrak{t}^1_{\mathbb{C}} = \mathbb{C}(H_{\alpha_2} + 2H_{\alpha_4} + H_{\alpha_5} + H_{\alpha_7})$$

---

63. 0220220 $\mathfrak{k}_{\mathbb{C}}^{(x,e,f)} \simeq T_1$

$$x = 8H_{\alpha_1} + 8H_{\alpha_2} + 16H_{\alpha_3} + 22H_{\alpha_4} + 18H_{\alpha_5} + 14H_{\alpha_6} + 8H_{\alpha_7}$$

$$e = \sqrt{5}X_{\alpha_1+\alpha_2+\alpha_3+2\alpha_4+\alpha_5+\alpha_6} + \sqrt{8}(X_{-\alpha_2-\alpha_4-\alpha_5-\alpha_6} + X_{\alpha_1+\alpha_2+\alpha_3+\alpha_4+\alpha_5+\alpha_6+\alpha_7}) + 3X_{\alpha_2+\alpha_3+2\alpha_4+2\alpha_5+\alpha_6} + X_{-\alpha_2-\alpha_3-\alpha_4} + \sqrt{5}X_{-\alpha_1-\alpha_2-\alpha_3-\alpha_4-\alpha_5}$$

$$\mathfrak{t}^1_{\mathbb{C}} = \mathbb{C}(H_{\alpha_2} + 2H_{\alpha_4} + H_{\alpha_5} + H_{\alpha_7})$$

---

66. 2204022 $\mathfrak{k}_{\mathbb{C}}^{(x,e,f)} \simeq T_2$

$$x = 12H_{\alpha_1} + 12H_{\alpha_2} + 22H_{\alpha_3} + 30H_{\alpha_4} + 26H_{\alpha_5} + 18H_{\alpha_6} + 10H_{\alpha_7}$$

$$e = \sqrt{8}X_{\alpha_1+\alpha_2+\alpha_3+2\alpha_4+\alpha_5} + \sqrt{18}X_{\alpha_1+\alpha_2+\alpha_3+\alpha_4+\alpha_5+\alpha_6} + \sqrt{10}X_{\alpha_2+\alpha_3+2\alpha_4+\alpha_5+\alpha_6+\alpha_7} + \sqrt{14}X_{-\alpha_1-\alpha_2-\alpha_3-\alpha_4} + \sqrt{10}X_{-\alpha_2-\alpha_4-\alpha_5-\alpha_6}$$

$$\mathfrak{t}^1_{\mathbb{C}} = \mathbb{C}(H_{\alpha_2} + H_{\alpha_3} + H_{\alpha_4} + H_{\alpha_5}) \oplus \mathbb{C}(H_{\alpha_3} + H_{\alpha_4} - H_{\alpha_7})$$

---

71. 2220222 $\mathfrak{k}_{\mathbb{C}}^{(x,e,f)} \simeq T_1$

$$x = 12H_{\alpha_1} + 12H_{\alpha_2} + 22H_{\alpha_3} + 30H_{\alpha_4} + 24H_{\alpha_5} + 18H_{\alpha_6} + 10H_{\alpha_7}$$

$$e = \sqrt{6}X_{\alpha_1+\alpha_2+\alpha_3+2\alpha_4+\alpha_5} + \sqrt{6}X_{\alpha_1+\alpha_2+\alpha_3+\alpha_4+\alpha_5+\alpha_6} + \sqrt{10}X_{\alpha_2+\alpha_3+\alpha_4+\alpha_5+\alpha_6+\alpha_7} + \sqrt{12}X_{\alpha_2+\alpha_3+2\alpha_4+2\alpha_5+\alpha_6} + \sqrt{12}X_{-\alpha_2-\alpha_3-\alpha_4-\alpha_5} + \sqrt{10}X_{-\alpha_2-\alpha_4-\alpha_5-\alpha_6}$$

$$\mathfrak{t}^1_{\mathbb{C}} = \mathbb{C}(H_{\alpha_2} + 2H_{\alpha_3} + 2H_{\alpha_4} - H_{\alpha_5} - H_{\alpha_7})$$

---

80. 4220224 $\mathfrak{k}_{\mathbb{C}}^{(x,e,f)} \simeq T_1$

$$x = 16H_{\alpha_1} + 16H_{\alpha_2} + 28H_{\alpha_3} + 38H_{\alpha_4} + 30H_{\alpha_5} + 22H_{\alpha_6} + 12H_{\alpha_7}$$

$$e = \sqrt{8+u}X_{\alpha_1+\alpha_2+\alpha_3+\alpha_4} + \sqrt{8-u}X_{\alpha_1+\alpha_2+\alpha_3+\alpha_4+\alpha_5} + \sqrt{22-u}X_{\alpha_2+\alpha_3+2\alpha_4+\alpha_5+\alpha_6} + \sqrt{12}X_{-\alpha_2-\alpha_4-\alpha_5-\alpha_6} + \sqrt{24}X_{-\alpha_2-\alpha_3-\alpha_4} + \sqrt{u}X_{\alpha_2+\alpha_3+2\alpha_4+2\alpha_5+\alpha_6} + \sqrt{12}X_{\alpha_2+\alpha_3+\alpha_4+\alpha_5+\alpha_6+\alpha_7}$$

$$\mathfrak{t}^1_{\mathbb{C}} = \mathbb{C}(H_{\alpha_2} + 2H_{\alpha_4} + H_{\alpha_5} + H_{\alpha_7})$$

---

84. 4224224 $\mathfrak{k}_{\mathbb{C}}^{(x,e,f)} \simeq T_1$

$$x = 20H_{\alpha_1} + 20H_{\alpha_2} + 36H_{\alpha_3} + 50H_{\alpha_4} + 42H_{\alpha_5} + 30H_{\alpha_6} + 16H_{\alpha_7}$$

$$e = \sqrt{42}X_{\alpha_1+\alpha_2+\alpha_3+\alpha_4+\alpha_5} + \sqrt{30}X_{\alpha_2+\alpha_3+2\alpha_4+\alpha_5+\alpha_6} + 4X_{\alpha_2+\alpha_3+\alpha_4+\alpha_5+\alpha_6+\alpha_7} + \sqrt{22}X_{-\alpha_1-\alpha_2-\alpha_3-\alpha_4} + \sqrt{30}X_{-\alpha_2-\alpha_3-\alpha_4-\alpha_5} + 4X_{-\alpha_2-\alpha_4-\alpha_5-\alpha_6}$$

$$\mathfrak{t}^1_{\mathbb{C}} = \mathbb{C}(H_{\alpha_2} + 2H_{\alpha_4} + H_{\alpha_5} + H_{\alpha_7})$$

EVI.

Let $\Delta = \{\alpha_1, \alpha_2, \ldots, \alpha_7\}$ be the Bourbaki simple roots of $\mathfrak{g}_{\mathbb{C}}$ then $\Delta_k = \{\beta_1, \ldots, \beta_7\}$, where $\beta_1 = \alpha_7$, $\beta_2 = \alpha_6$, $\beta_3 = \alpha_5$, $\beta_4 = \alpha_4$, $\beta_5 = \alpha_3$, $\beta_6 = \alpha_2$ and $\beta_7 = 2\alpha_1+2\alpha_2+3\alpha_3+4\alpha_4+3\alpha_5+2\alpha_6+\alpha_7$, is a set of simple roots for $\mathfrak{k}_{\mathbb{C}} = \mathfrak{so}_{12}(\mathbb{C})\oplus\mathfrak{sl}_2(\mathbb{C})$.

Table IX

---

6. 000000 4 $\mathfrak{k}_{\mathbb{C}}^{(x,e,f)} \simeq A_5$

$$x = 4H_{\alpha_1} + 4H_{\alpha_2} + 6H_{\alpha_3} + 8H_{\alpha_4} + 6H_{\alpha_5} + 4H_{\alpha_6} + 2H_{\alpha_7}$$

$$e = \sqrt{2}(X_{\alpha_1+\alpha_2+2\alpha_3+2\alpha_4+2\alpha_5+2\alpha_6+\alpha_7} + X_{\alpha_1+\alpha_2+\alpha_3+2\alpha_4+\alpha_5})$$

$$\mathfrak{t}^1_{\mathbb{C}} = \mathbb{C}H_{\alpha_2} \oplus \mathbb{C}H_{\alpha_5} \oplus \mathbb{C}H_{\alpha_7} \oplus \mathbb{C}(H_{\alpha_3} + H_{\alpha_4}) \oplus \mathbb{C}(H_{\alpha_4} + H_{\alpha_6})$$

---

7. 000020 2 $\mathfrak{k}_{\mathbb{C}}^{(x,e,f)} \simeq C_3$

$$x = 2H_{\alpha_1} + 4H_{\alpha_2} + 6H_{\alpha_3} + 8H_{\alpha_4} + 6H_{\alpha_5} + 4H_{\alpha_6} + 2H_{\alpha_7}$$

$$e = X_{\alpha_1+\alpha_2+2\alpha_3+2\alpha_4+\alpha_5} + X_{\alpha_1+\alpha_2+2\alpha_3+2\alpha_4+2\alpha_5+2\alpha_6+\alpha_7} + X_{\alpha_1+2\alpha_2+2\alpha_3+4\alpha_4+3\alpha_5+2\alpha_6+\alpha_7} + X_{-\alpha_1}$$

$$\mathfrak{t}^1_{\mathbb{C}} = \mathbb{C}H_{\alpha_2} \oplus \mathbb{C}H_{\alpha_5} \oplus \mathbb{C}H_{\alpha_7}$$

---

8. 020000 0 $\mathfrak{k}_{\mathbb{C}}^{(x,e,f)} \simeq A_3 + A_1 + T_1$

$x = 2H_{\alpha_2} + 2H_{\alpha_3} + 4H_{\alpha_4} + 4H_{\alpha_5} + 4H_{\alpha_6} + 2H_{\alpha_7}$

$e = \sqrt{2}(X_{\alpha_1+\alpha_2+\alpha_3+2\alpha_4+2\alpha_5+2\alpha_6+\alpha_7} + X_{-\alpha_1})$

$\mathfrak{t}_{\mathbb{C}}^1 = \mathbb{C}H_{\alpha_2} \oplus \mathbb{C}H_{\alpha_4} \oplus \mathbb{C}H_{\alpha_5} \oplus \mathbb{C}H_{\alpha_7} \oplus \mathbb{C}(H_{\alpha_1} + 2H_{\alpha_3} + H_{\alpha_6})$

---

14. 400000 0 $\mathfrak{k}_{\mathbb{C}}^{(x,e,f)} \simeq G_2 + A_1$

$x = 2H_{\alpha_2} + 2H_{\alpha_3} + 4H_{\alpha_4} + 4H_{\alpha_5} + 4H_{\alpha_6} + 4H_{\alpha_7}$

$e = \sqrt{2}(X_{\alpha_1+\alpha_3+\alpha_4+\alpha_5+\alpha_6+\alpha_7} + X_{\alpha_1+2\alpha_2+2\alpha_3+3\alpha_4+2\alpha_5+\alpha_6+\alpha_7})$
$+\sqrt{2}(X_{-\alpha_1} + X_{-\alpha_1-\alpha_2-2\alpha_3-2\alpha_4-\alpha_5})$

$\mathfrak{t}_{\mathbb{C}}^1 = \mathbb{C}H_{\alpha_4} \oplus \mathbb{C}H_{\alpha_5} \oplus \mathbb{C}(2H_{\alpha_1} + 3H_{\alpha_2} + 4H_{\alpha_3} + 4H_{\alpha_6} + H_{\alpha_7})$

---

15. 000200 0 $\mathfrak{k}_{\mathbb{C}}^{(x,e,f)} \simeq 2A_1 + T_1$

$x = 4H_{\alpha_2} + 4H_{\alpha_3} + 8H_{\alpha_4} + 6H_{\alpha_5} + 4H_{\alpha_6} + 2H_{\alpha_7}$

$e = \sqrt{2}(X_{\alpha_1+\alpha_2+2\alpha_3+3\alpha_4+2\alpha_5+\alpha_6} + X_{\alpha_1+2\alpha_2+2\alpha_3+3\alpha_4+2\alpha_5+\alpha_6+\alpha_7})$
$+\sqrt{2}(X_{-\alpha_1-\alpha_3-\alpha_4} + X_{-\alpha_1-\alpha_2-\alpha_3-\alpha_4-\alpha_5})$

$\mathfrak{t}_{\mathbb{C}}^1 = \mathbb{C}H_{\alpha_3} \oplus \mathbb{C}(H_{\alpha_1} + 2H_{\alpha_2} + 2H_{\alpha_4} + H_{\alpha_5} + 2H_{\alpha_6}) \oplus \mathbb{C}(-H_{\alpha_2} + H_{\alpha_5} + H_{\alpha_7})$

---

19. 000040 0 $\mathfrak{k}_{\mathbb{C}}^{(x,e,f)} \simeq 3A_1$

$x = 4H_{\alpha_2} + 6H_{\alpha_3} + 8H_{\alpha_4} + 6H_{\alpha_5} + 4H_{\alpha_6} + 2H_{\alpha_7}$

$e = 2X_{\alpha_1+\alpha_2+2\alpha_3+2\alpha_4+\alpha_5} + \sqrt{2-u}X_{\alpha_1+\alpha_2+2\alpha_3+2\alpha_4+2\alpha_5+2\alpha_6+\alpha_7}$
$+\sqrt{4-u}X_{-\alpha_1-\alpha_3}+\sqrt{2+u}X_{-\alpha_1-\alpha_2-\alpha_3-2\alpha_4-\alpha_5}+\sqrt{u}X_{\alpha_1+2\alpha_2+2\alpha_3+4\alpha_4+3\alpha_5+2\alpha_6+\alpha_7}$

$\mathfrak{t}_{\mathbb{C}}^1 = \mathbb{C}H_{\alpha_2} \oplus \mathbb{C}H_{\alpha_5} \oplus \mathbb{C}H_{\alpha_7}$

---

20. 000200 4 $\mathfrak{k}_{\mathbb{C}}^{(x,e,f)} \simeq 3A_1$

$x = 4H\alpha_1 + 8H_{\alpha_2} + 10H_{\alpha_3} + 16H_{\alpha_4} + 12H_{\alpha_5} + 8H_{\alpha_6} + 4H_{\alpha_7}$

$e = \sqrt{2-u}X_{\alpha_1+\alpha_2+\alpha_3+2\alpha_4+\alpha_5} + \sqrt{2+u}X_{\alpha_1+\alpha_2+2\alpha_3+2\alpha_4+\alpha_5}$
$+2X_{\alpha_1+\alpha_2+\alpha_3+2\alpha_4+2\alpha_5+2\alpha_6+\alpha_7} + \sqrt{4-u}X_{-\alpha_1} + \sqrt{u}X_{-\alpha_1-\alpha_3}$

$\mathfrak{t}_{\mathbb{C}}^1 = \mathbb{C}H_{\alpha_2} \oplus \mathbb{C}H_{\alpha_5} \oplus \mathbb{C}H_{\alpha_7}$

---

21. 020020 2 $\mathfrak{k}_{\mathbb{C}}^{(x,e,f)} \simeq 2A_1$

$x = 2H_{\alpha_1} + 6H_{\alpha_2} + 8H_{\alpha_3} + 12H_{\alpha_4} + 10H_{\alpha_5} + 8H_{\alpha_6} + 4H_{\alpha_7}$

$e = \sqrt{3}(X_{\alpha_1+\alpha_2+2\alpha_3+2\alpha_4+2\alpha_5+\alpha_6+\alpha_7} + X_{\alpha_1+\alpha_2+2\alpha_3+2\alpha_4+\alpha_5+\alpha_6})$
$+2X_{-\alpha_1-\alpha_3} + X_{\alpha_1+\alpha_2+\alpha_3+2\alpha_4+2\alpha_5+2\alpha_6+\alpha_7} + X_{-\alpha_1-\alpha_2-\alpha_3-2\alpha_4-\alpha_5}$

$\mathfrak{t}_{\mathbb{C}}^1 = \mathbb{C}H_{\alpha_2} \oplus \mathbb{C}(H_{\alpha_5} - H_{\alpha_7})$

---

22. 000040 8 $\mathfrak{k}_{\mathbb{C}}^{(x,e,f)} \simeq C_3$

$x = 8H_{\alpha_1} + 12H_{\alpha_2} + 18H_{\alpha_3} + 24H_{\alpha_4} + 18H_{\alpha_5} + 12H_{\alpha_6} + 6H_{\alpha_7}$

$e = \sqrt{6}(X_{\alpha_1+\alpha_2+\alpha_3+\alpha_4+\alpha_5} + X_{\alpha_1+\alpha_3+\alpha_4+\alpha_5+\alpha_6+\alpha_7})$

$$+\sqrt{10}X_{-\alpha_1} + \sqrt{6}X_{\alpha_1+\alpha_2+\alpha_3+2\alpha_4+\alpha_5+\alpha_6}$$
$$\mathfrak{t}^1_{\mathbb{C}} = \mathbb{C}(H_{\alpha_4} + H_{\alpha_5}) \oplus \mathbb{C}(H_{\alpha_5} + H_{\alpha_6}) \oplus \mathbb{C}(H_{\alpha_2} + H_{\alpha_6} + H_{\alpha_7})$$

---

23. 020040 4 $\mathfrak{k}_{\mathbb{C}}^{(x,e,f)} \simeq B2 + A_1$

$$x = 4H_{\alpha_1} + 10H_{\alpha_2} + 14H_{\alpha_3} + 20H_{\alpha_4} + 16H_{\alpha_5} + 12H_{\alpha_6} + 6H_{\alpha_7}$$
$$e = \sqrt{10}(X_{\alpha_1+\alpha_2+2\alpha_3+2\alpha_4+\alpha_5} + \sqrt{6}X_{\alpha_1+\alpha_2+\alpha_3+2\alpha_4+2\alpha_5+2\alpha_6+\alpha_7}$$
$$+\sqrt{6}X_{-\alpha_1-\alpha_3} + X_{-\alpha_1-\alpha_2-\alpha_3-2\alpha_4-\alpha_5})$$
$$\mathfrak{t}^1_{\mathbb{C}} = \mathbb{C}H_{\alpha_2} \oplus \mathbb{C}H_{\alpha_5} \oplus \mathbb{C}H_{\alpha_7}$$

---

25. 040000 4 $\mathfrak{k}_{\mathbb{C}}^{(x,e,f)} \simeq A_2 + T_1$

$$x = 4H_{\alpha_1} + 8H_{\alpha_2} + 10H_{\alpha_3} + 16H_{\alpha_4} + 14H_{\alpha_5} + 12H_{\alpha_6} + 6H_{\alpha_7}$$
$$e = \sqrt{6}(X_{\alpha_1+\alpha_3+\alpha_4+\alpha_5+\alpha_6} + X_{\alpha_1+2\alpha_2+2\alpha_3+3\alpha_4+2\alpha_5+\alpha_6+\alpha_7})$$
$$+2(X_{-\alpha_1-\alpha_3-\alpha_4} + X_{-\alpha_1-\alpha_2-\alpha_3-\alpha_4-\alpha_5})$$
$$\mathfrak{t}^1_{\mathbb{C}} = \mathbb{C}H_{\alpha_3} \oplus \mathbb{C}(H_{\alpha_4} + H_{\alpha_5}) \oplus \mathbb{C}(H_{\alpha_2} + H_{\alpha_4} - H_{\alpha_7})$$

---

26. 020200 0 $\mathfrak{k}_{\mathbb{C}}^{(x,e,f)} \simeq A_1 + T_2$

$$x = 6H_{\alpha_2} + 6H_{\alpha_3} + 12H_{\alpha_4} + 10H_{\alpha_5} + 8H_{\alpha_6} + 4H_{\alpha_7}$$
$$e = 2X_{\alpha_1+\alpha_2+\alpha_3+2\alpha_4+2\alpha_5+2\alpha_6+\alpha_7} + \sqrt{6}X_{\alpha_1+\alpha_2+2\alpha_3+3\alpha_4+2\alpha_5+\alpha_6}$$
$$+2X_{-\alpha_1-\alpha_2-\alpha_3-2\alpha_4-\alpha_5} + \sqrt{6}X_{-\alpha_1-\alpha_3-\alpha_4-\alpha_5-\alpha_6}$$
$$\mathfrak{t}^1_{\mathbb{C}} = \mathbb{C}H_{\alpha_5} \oplus \mathbb{C}(H_{\alpha_2} - H_{\alpha_7}) \oplus \mathbb{C}(H_{\alpha_1} + 2H_{\alpha_2} + 2H_{\alpha_3} + 2H_{\alpha_4} + H_{\alpha_6})$$

---

29. 004000 0 $\mathfrak{k}_{\mathbb{C}}^{(x,e,f)} \simeq A_1$

$$x = 6H_{\alpha_2} + 6H_{\alpha_3} + 12H_{\alpha_4} + 12H_{\alpha_5} + 8H_{\alpha_6} + 4H_{\alpha_7}$$
$$e = \sqrt{6}X_{\alpha_1+\alpha_2+\alpha_3+2\alpha_4+2\alpha_5+\alpha_6} + 2X_{\alpha_1+\alpha_2+2\alpha_3+2\alpha_4+2\alpha_5+\alpha_6+\alpha_7}$$
$$+\sqrt{2}X_{\alpha_1+2\alpha_2+2\alpha_3+3\alpha_4+2\alpha_5+2\alpha_6+\alpha_7} + \sqrt{6}X_{-\alpha_1-\alpha_2-\alpha_3-\alpha_4-\alpha_5}$$
$$+2X_{-\alpha_1-\alpha_3-\alpha_4-\alpha_5-\alpha_6} + \sqrt{2}X_{-\alpha_1-\alpha_2-\alpha_3-2\alpha_4-\alpha_5-\alpha_6-\alpha_7}$$
$$\mathfrak{t}^1_{\mathbb{C}} = \mathbb{C}(2H_{\alpha_1} + 3H_{\alpha_2} + 6H_{\alpha_3} + 8H_{\alpha_4} + 3H_{\alpha_5} - H_{\alpha_7})$$

---

31. 020220 2 $\mathfrak{k}_{\mathbb{C}}^{(x,e,f)} \simeq A_1$

$$x = 2H_{\alpha_1} + 10H_{\alpha_2} + 12H_{\alpha_3} + 20H_{\alpha_4} + 16H_{\alpha_5} + 12H_{\alpha_6} + 6H_{\alpha_7}$$
$$e = \sqrt{5}(X_{\alpha_1+\alpha_2+2\alpha_3+2\alpha_4+\alpha_5+\alpha_6} + X_{\alpha_1+\alpha_2+2\alpha_3+2\alpha_4+2\alpha_5+\alpha_6+\alpha_7})$$
$$+3X_{\alpha_1+\alpha_2+\alpha_3+2\alpha_4+2\alpha_5+2\alpha_6+\alpha_7} + X_{-\alpha_1-\alpha_2-\alpha_3-2\alpha_4-\alpha_5}$$
$$+\sqrt{8}(X_{-\alpha_1-\alpha_3-\alpha_4-\alpha_5-\alpha_6} + X_{-\alpha_1-\alpha_2-\alpha_3-\alpha_4-\alpha_5-\alpha_6-\alpha_7})$$
$$\mathfrak{t}^1_{\mathbb{C}} = \mathbb{C}(H_{\alpha_2} + H_{\alpha_5} - H_{\alpha_7})$$

---

32. 000400 4 $\mathfrak{k}_{\mathbb{C}}^{(x,e,f)} \simeq A_1$

$$x = 4H_{\alpha_1} + 12H_{\alpha_2} + 14H_{\alpha_3} + 24H_{\alpha_4} + 18H_{\alpha_5} + 12H_{\alpha_6} + 6H_{\alpha_7}$$
$$e = \sqrt{5}(X_{\alpha_1+\alpha_2+\alpha_3+2\alpha_4+\alpha_5+\alpha_6} + X_{\alpha_1+\alpha_2+\alpha_3+2\alpha_4+2\alpha_5+\alpha_6+\alpha_7}$$
$$+3X_{\alpha_1+\alpha_2+2\alpha_3+2\alpha_4+2\alpha_5+2\alpha_6+\alpha_7} + \sqrt{8}(X_{-\alpha_1-\alpha_2-\alpha_3-\alpha_4-\alpha_5-\alpha_6}$$

$+X_{-\alpha_1-\alpha_3-\alpha_4-\alpha_5-\alpha_6-\alpha_7}) + X_{\alpha_1+\alpha_2+2\alpha_3+2\alpha_4+\alpha_5}$

$\mathfrak{t}^1_{\mathbb{C}} = \mathbb{C}(H_{\alpha_2} - H_{\alpha_5} + H_{\alpha_7})$

---

33. 020240 4 $\quad \mathfrak{k}_{\mathbb{C}}^{(x,e,f)} \simeq 2A_1$

$x = 4H_{\alpha_1} + 14H_{\alpha_2} + 18H_{\alpha_3} + 28H_{\alpha_4} + 22H_{\alpha_5} + 16H_{\alpha_6} + 8H_{\alpha_7}$

$$e = \sqrt{18}X_{\alpha_1+\alpha_2+\alpha_3+2\alpha_4+2\alpha_5+2\alpha_6+\alpha_7} + \sqrt{8}X_{-\alpha_1-\alpha_2-\alpha_3-2\alpha_4-\alpha_5} +\sqrt{10}(X_{-\alpha_1-\alpha_3-\alpha_4-\alpha_5-\alpha_6} + X_{-\alpha_1-\alpha_2-\alpha_3-\alpha_4-\alpha_5-\alpha_6-\alpha_7}) +\sqrt{14}X_{\alpha_1+\alpha_2+2\alpha_3+2\alpha_4+\alpha_5}$$

$\mathfrak{t}^1_{\mathbb{C}} = \mathbb{C}H_{\alpha_5} \oplus \mathbb{C}(H_{\alpha_2} - H_{\alpha_7})$

---

34. 040040 8 $\quad \mathfrak{k}_{\mathbb{C}}^{(x,e,f)} \simeq 2A_1$

$x = 8H_{\alpha_1} + 16H_{\alpha_2} + 22H_{\alpha_3} + 32H_{\alpha_4} + 26H_{\alpha_5} + 20H_{\alpha_6} + 10H_{\alpha_7}$

$$e = \sqrt{10}(X_{\alpha_1+\alpha_2+\alpha_3+\alpha_4+\alpha_5+\alpha_6+\alpha_7} + X_{\alpha_1+\alpha_2+\alpha_3+2\alpha_4+2\alpha_5+\alpha_6}) +\sqrt{8}X_{-\alpha_1-\alpha_3-\alpha_4} + \sqrt{18}X_{-\alpha_1-\alpha_2-\alpha_3-\alpha_4-\alpha_5} +\sqrt{14}X_{\alpha_1+\alpha_2+2\alpha_3+2\alpha_4+\alpha_5}$$

$\mathfrak{t}^1_{\mathbb{C}} = \mathbb{C}(H_{\alpha_2} + H_{\alpha_4}) \oplus \mathbb{C}(H_{\alpha_2} - H_{\alpha_5} - H_{\alpha_7})$

---

35. 400400 0 $\quad \mathfrak{k}_{\mathbb{C}}^{(x,e,f)} \simeq A_1$

$x = 10H_{\alpha_2} + 10H_{\alpha_3} + 20H_{\alpha_4} + 16H_{\alpha_5} + 12H_{\alpha_6} + 8H_{\alpha_7}$

$$e = \sqrt{6}X_{\alpha_1+\alpha_2+\alpha_3+2\alpha_4+\alpha_5+\alpha_6+\alpha_7} + \sqrt{10}X_{\alpha_1+\alpha_2+2\alpha_3+3\alpha_4+2\alpha_5+\alpha_6} +\sqrt{10}X_{-\alpha_1-\alpha_3-\alpha_4-\alpha_5-\alpha_6-\alpha_7} + \sqrt{12}X_{-\alpha_1-\alpha_2-2\alpha_3-2\alpha_4-\alpha_5} +\sqrt{12}X_{\alpha_1+\alpha_2+2\alpha_3+2\alpha_4+2\alpha_5+\alpha_6+\alpha_7} + \sqrt{6}X_{-\alpha_1-\alpha_2-\alpha_3-2\alpha_4-2\alpha_5-\alpha_6}$$

$\mathfrak{t}^1_{\mathbb{C}} = \mathbb{C}(2H_{\alpha_1} + 3H_{\alpha_2} + 4H_{\alpha_3} + 4H_{\alpha_4} + 3H_{\alpha_5} + 4H_{\alpha_6} + H_{\alpha_7})$

---

36. 040400 4 $\quad \mathfrak{k}_{\mathbb{C}}^{(x,e,f)} \simeq T_1$

$x = 4H_{\alpha_1} + 16H_{\alpha_2} + 18H_{\alpha_3} + 32H_{\alpha_4} + 26H_{\alpha_5} + 20H_{\alpha_6} + 10H_{\alpha_7}$

$$e = \sqrt{22}X_{\alpha_1+\alpha_2+\alpha_3+2\alpha_4+\alpha_5+\alpha_6} + \sqrt{22-u}X_{\alpha_1+\alpha_2+2\alpha_3+2\alpha_4+2\alpha_5+\alpha_6+\alpha_7} +\sqrt{12}X_{-\alpha_1-\alpha_2-\alpha_3-\alpha_4-\alpha_5-\alpha_6} + \sqrt{8+u}X_{-\alpha_1-\alpha_2-\alpha_3-2\alpha_4-\alpha_5} +\sqrt{u}X_{\alpha_1+\alpha_2+\alpha_3+2\alpha_4+2\alpha_5+\alpha_6+\alpha_7} + \sqrt{12}X_{-\alpha_1-\alpha_3-\alpha_4-\alpha_5-\alpha_6-\alpha_7} +\sqrt{8+u}X_{-\alpha_1-\alpha_2-2\alpha_3-2\alpha_4-\alpha_5}$$

$\mathfrak{t}^1_{\mathbb{C}} = \mathbb{C}(H_{\alpha_2} - H_{\alpha_5} + H_{\alpha_7})$

---

37. 040440 8 $\quad \mathfrak{k}_{\mathbb{C}}^{(x,e,f)} \simeq A_1$

$x = 8H_{\alpha_1} + 24H_{\alpha_2} + 30H_{\alpha_3} + 48H_{\alpha_4} + 38H_{\alpha_5} + 28H_{\alpha_6} + 14H_{\alpha_7}$

$$e = \sqrt{22}X_{\alpha_1+\alpha_2+2\alpha_3+2\alpha_4+\alpha_5} + \sqrt{30}X_{\alpha_1+\alpha_2+\alpha_3+2\alpha_4+\alpha_5+\alpha_6} +4X_{-\alpha_1-\alpha_2-\alpha_3-\alpha_4-\alpha_5-\alpha_6} + \sqrt{42}X_{-\alpha_1-\alpha_2-\alpha_3-2\alpha_4-\alpha_5} +\sqrt{30}X_{\alpha_1+\alpha_2+\alpha_3+2\alpha_4+2\alpha_5+\alpha_6+\alpha_7} + 4X_{-\alpha_1-\alpha_3-\alpha_4-\alpha_5-\alpha_6-\alpha_7}$$

$\mathfrak{t}^1_{\mathbb{C}} = \mathbb{C}(H_{\alpha_2} - H_{\alpha_5} + H_{\alpha_7})$

EVII.

Let $\Delta = \{\alpha_1, \alpha_2, \ldots, \alpha_7\}$ be the Bourbaki simple roots of $\mathfrak{g}_{\mathbb{C}}$ then $\Delta_k = \{\beta_1, \ldots, \beta_7\}$, where $\beta_1 = \alpha_6$, $\beta_2 = \alpha_2$, $\beta_3 = \alpha_5$, $\beta_4 = \alpha_4$, $\beta_5 = \alpha_3$, $\beta_6 = \alpha_1$ and $\beta_7 = -2\alpha_1 - 2\alpha_2 - 3\alpha_3 - 4\alpha_4 - 3\alpha_5 - 2\alpha_6 - \alpha_7$, is a set of simple roots for $\mathfrak{k}_{\mathbb{C}} = E_6 \oplus (\mathbb{C})$.

Table X

---

6. 000000 2 $\quad \mathfrak{k}_{\mathbb{C}}^{(x,e,f)} \simeq F_4$

$$x = -(2H_{\alpha_1} + 3H_{\alpha_2} + 4H_{\alpha_3} + 6H_{\alpha_4} + 5H_{\alpha_5} + 4H_{\alpha_6} + 3H_{\alpha_7})$$
$$e = X_{-\alpha_1} + X_{-\alpha_2-\alpha_3-2\alpha_4-2\alpha_5-2\alpha_6-\alpha_7} + X_{-2\alpha_1-2\alpha_2-3\alpha_3-4\alpha_4-3\alpha_5-2\alpha_6-\alpha_7}$$
$$\mathfrak{t}_{\mathbb{C}}^1 = \mathbb{C}H_{\alpha_2} \oplus \mathbb{C}H_{\alpha_3} \oplus \mathbb{C}H_{\alpha_4} \oplus \mathbb{C}H_{\alpha_5}$$

---

7. 000000 -2 $\quad \mathfrak{k}_{\mathbb{C}}^{(x,e,f)} \simeq F_4$

$$x = 2H_{\alpha_1} + 3H_{\alpha_2} + 4H_{\alpha_3} + 6H_{\alpha_4} + 5H_{\alpha_5} + 4H_{\alpha_6} + 3H_{\alpha_7}$$
$$e = X_{\alpha_1} + X_{\alpha_2+\alpha_3+2\alpha_4+2\alpha_5+2\alpha_6+\alpha_7} + X_{2\alpha_1+2\alpha_2+3\alpha_3+4\alpha_4+3\alpha_5+2\alpha_6+\alpha_7}$$
$$\mathfrak{t}_{\mathbb{C}}^1 = \mathbb{C}H_{\alpha_2} \oplus \mathbb{C}H_{\alpha_3} \oplus \mathbb{C}H_{\alpha_4} \oplus \mathbb{C}H_{\alpha_5}$$

---

8. 000002 -2 $\quad \mathfrak{k}_{\mathbb{C}}^{(x,e,f)} \simeq B_4$

$$x = 2H_{\alpha_1} + H_{\alpha_2} + 2H_{\alpha_3} + 2H_{\alpha_4} + H_{\alpha_5} + -H_{\alpha_7}$$
$$e = X_{-\alpha_7} + X_{-(\alpha_2+\alpha_3+2\alpha_4+2\alpha_5+2\alpha_6+\alpha_7)} + X_{2\alpha_1+2\alpha_2+3\alpha_3+4\alpha_4+3\alpha_5+2\alpha_6+\alpha_7}$$
$$\mathfrak{t}_{\mathbb{C}}^1 = \mathbb{C}H_{\alpha_2} \oplus \mathbb{C}H_{\alpha_3} \oplus \mathbb{C}H_{\alpha_4} \oplus \mathbb{C}H_{\alpha_5}$$

---

9. 200000 -2 $\quad \mathfrak{k}_{\mathbb{C}}^{(x,e,f)} \simeq b_4$

$$x = 2H_{\alpha_1} + 3H_{\alpha_2} + 4H_{\alpha_3} + 6H_{\alpha_4} + 5H_{\alpha_5} + 4H_{\alpha_6} + H_{\alpha_7}$$
$$e = X_{-\alpha_7} + X_{\alpha_2+\alpha_3+2\alpha_4+2\alpha_5+2\alpha_6+\alpha_7} + X_{2\alpha_1+2\alpha_2+3\alpha_3+4\alpha_4+3\alpha_5+2\alpha_6+\alpha_7}$$
$$\mathfrak{t}_{\mathbb{C}}^1 = \mathbb{C}H_{\alpha_2} \oplus \mathbb{C}H_{\alpha_3} \oplus \mathbb{C}H_{\alpha_4} \oplus \mathbb{C}H_{\alpha_5}$$

---

10. 020000 -2 $\quad \mathfrak{k}_{\mathbb{C}}^{(x,e,f)} \simeq A_4 + T_1$

$$x = 2H_{\alpha_1} + 4H_{\alpha_2} + 4H_{\alpha_3} + 6H_{\alpha_4} + 4H_{\alpha_5} + 2H_{\alpha_6}$$
$$e = \sqrt{2}(X_{\alpha_1+2\alpha_2+2\alpha_3+3\alpha_4+2\alpha_5+\alpha_6+\alpha_7} + X_{-\alpha_7})$$
$$\mathfrak{t}_{\mathbb{C}}^1 = \mathbb{C}H_{\alpha_1} \oplus \mathbb{C}H_{\alpha_3} \oplus \mathbb{C}H_{\alpha_4} \oplus \mathbb{C}H_{\alpha_5} \oplus \mathbb{C}(H_{\alpha_2} + 2H_{\alpha_6} + H_{\alpha_7})$$

---

15. 200002 -4 $\quad \mathfrak{k}_{\mathbb{C}}^{(x,e,f)} \simeq G_2 + T_1$

$$x = 4H_{\alpha_1} + 4H_{\alpha_2} + 6H_{\alpha_3} + 8H_{\alpha_4} + 6H_{\alpha_5} + 4H_{\alpha_6}$$
$$e = \sqrt{2}(X_{\alpha_1+\alpha_2+\alpha_3+2\alpha_4+2\alpha_5+2\alpha_6+\alpha_7} + X_{-\alpha_6-\alpha_7} + X_{-\alpha_2-\alpha_3-2\alpha_4-2\alpha_5-\alpha_6-\alpha_7} + X_{\alpha_1+2\alpha_2+3\alpha_3+4\alpha_4+3\alpha_5+2\alpha_6+\alpha_7})$$

$\mathfrak{t}^1_{\mathbb{C}} = \mathbb{C}H_{\alpha_2} \oplus \mathbb{C}H_{\alpha_4} \oplus \mathbb{C}(2H_{\alpha_1} + 2H_{\alpha_3} + H_{\alpha_5} + H_{\alpha_7})$

---

16. 200002 -2 $\quad \mathfrak{k}^{(x,e,f)}_{\mathbb{C}} \simeq B_3$

$x = 2H_{\alpha_1} + H_{\alpha_2} + 2H_{\alpha_3} + 2H_{\alpha_4} + H_{\alpha_5} - 3H_{\alpha_7}$

$e = 2X_{2\alpha_1+2\alpha_2+3\alpha_3+4\alpha_4+3\alpha_5+2\alpha_6+\alpha_7} + \sqrt{3}(X_{-\alpha_1-\alpha_2-2\alpha_3-2\alpha_4-\alpha_5-\alpha_6-\alpha_7} + X_{-\alpha_1-\alpha_2-\alpha_3-2\alpha_4-2\alpha_5-\alpha_6-\alpha_7}) + X_{-\alpha_2-\alpha_3-2\alpha_4-2\alpha_5-2\alpha_6-\alpha_7}$

$\mathfrak{t}^1_{\mathbb{C}} = \mathbb{C}H_{\alpha_2} \oplus \mathbb{C}H_{\alpha_4} \oplus \mathbb{C}(H_{\alpha_3} + H_{\alpha_5})$

---

17. 400000 -2 $\quad \mathfrak{k}^{(x,e,f)}_{\mathbb{C}} \simeq B_3$

$x = 2H_{\alpha_1} + 3H_{\alpha_2} + 4H_{\alpha_3} + 6H_{\alpha_4} + 5H_{\alpha_5} + 4H_{\alpha_6} - H_{\alpha_7}$

$e = 2X_{2\alpha_1+2\alpha_2+3\alpha_3+4\alpha_4+3\alpha_5+2\alpha_6+\alpha_7} + X_{\alpha_2+\alpha_3+2\alpha_4+2\alpha_5+2\alpha_6+\alpha_7} + \sqrt{3}(X_{-\alpha_1-\alpha_3-\alpha_4-\alpha_5-\alpha_6-\alpha_7} + X_{-\alpha_1-2\alpha_2-2\alpha_3-3\alpha_4-2\alpha_5-\alpha_6-\alpha_7})$

$\mathfrak{t}^1_{\mathbb{C}} = \mathbb{C}H_{\alpha_3} \oplus \mathbb{C}H_{\alpha_4} \oplus \mathbb{C}H_{\alpha_5}$

---

18. 000004 -6 $\quad \mathfrak{k}^{(x,e,f)}_{\mathbb{C}} \simeq B_3$

$x = 6H_{\alpha_1} + 5H_{\alpha_2} + 8H_{\alpha_3} + 10H_{\alpha_4} + 7H_{\alpha_5} + 4H_{\alpha_6} + H_{\alpha_7}$

$e = \sqrt{3}(X_{\alpha_1+2\alpha_2+2\alpha_3+3\alpha_4+2\alpha_5+\alpha_6+\alpha_7} + X_{\alpha_1+\alpha_3+\alpha_4+\alpha_5+\alpha_6+\alpha_7}) + 2X_{-\alpha_7} + X_{-\alpha_2-\alpha_3-2\alpha_4-2\alpha_5-2\alpha_6-\alpha_7}$

$\mathfrak{t}^1_{\mathbb{C}} = \mathbb{C}H_{\alpha_3} \oplus \mathbb{C}H_{\alpha_4} \oplus \mathbb{C}H_{\alpha_5}$

---

19. 200002 -6 $\quad \mathfrak{k}^{(x,e,f)}_{\mathbb{C}} \simeq B_3$

$x = 6H_{\alpha_1} + 7H_{\alpha_2} + 10H_{\alpha_3} + 14H_{\alpha_4} + 11H_{\alpha_5} + 8H_{\alpha_6} + 3H_{\alpha_7}$

$e = \sqrt{3}(X_{\alpha_1+\alpha_2+2\alpha_3+2\alpha_4+\alpha_5+\alpha_6+\alpha_7} + X_{\alpha_1+\alpha_2+\alpha_3+2\alpha_4+2\alpha_5+\alpha_6+\alpha_7}) + 2X_{-\alpha_7} + X_{\alpha_2+\alpha_3+2\alpha_4+2\alpha_5+2\alpha_6+\alpha_7}$

$\mathfrak{t}^1_{\mathbb{C}} = \mathbb{C}H_{\alpha_2} \oplus \mathbb{C}H_{\alpha_4} \oplus \mathbb{C}(H_{\alpha_3} + H_{\alpha_5})$

---

20. 220002 -6 $\quad \mathfrak{k}^{(x,e,f)}_{\mathbb{C}} \simeq A_2 + T1$

$x = 6H_{\alpha_1} + 8H_{\alpha_2} + 10H_{\alpha_3} + 14H_{\alpha_4} + 10H_{\alpha_5} + 6H_{\alpha_6}$

$e = \sqrt{6}(X_{\alpha_1+\alpha_2+2\alpha_3+2\alpha_4+2\alpha_5+2\alpha_6+\alpha_7} + X_{-\alpha_2-\alpha_3-\alpha_4-\alpha_5-\alpha_6-\alpha_7}) + 2(X_{\alpha_1+2\alpha_2+2\alpha_3+3\alpha_4+2\alpha_5+\alpha_6+\alpha_7} + X_{-\alpha_1-\alpha_3-\alpha_4-\alpha_5-\alpha_6-\alpha_7})$

$\mathfrak{t}^1_{\mathbb{C}} = \mathbb{C}H_{\alpha_5} \oplus \mathbb{C}(H_{\alpha_3} + H_{\alpha_4}) \oplus \mathbb{C}(H_{\alpha_2} - 2H_{\alpha_3} + 2H_{\alpha_6} + H_{\alpha_7})$

---

21. 400004 -6 $\quad \mathfrak{k}^{(x,e,f)}_{\mathbb{C}} \simeq G_2$

$x = 6H_{\alpha_1} + 5H_{\alpha_2} + 8H_{\alpha_3} + 10H_{\alpha_4} + 7H_{\alpha_5} + 4H_{\alpha_6} - 3H_{\alpha_7}$

$e = \sqrt{8}(X_{\alpha_1+\alpha_2+\alpha_3+2\alpha_4+2\alpha_5+2\alpha_6+\alpha_7} + X_{\alpha_1+2\alpha_2+3\alpha_3+4\alpha_4+3\alpha_5+2\alpha_6+\alpha_7}) + \sqrt{5}(X_{-\alpha_1-\alpha_2-2\alpha_3-2\alpha_4-\alpha_5-\alpha_6-\alpha_7} + X_{-\alpha_1-\alpha_2-\alpha_3-2\alpha_4-2\alpha_5-\alpha_6-\alpha_7}) + 3X_{-\alpha_2-\alpha_3-2\alpha_4-2\alpha_5-2\alpha_6-\alpha_7}$

$\mathfrak{t}^1_{\mathbb{C}} = \mathbb{C}H_{\alpha_2} \oplus \mathbb{C}H_{\alpha_4}$

---

22. 400004 -10 $\quad \mathfrak{k}_{\mathbb{C}}^{(x,e,f)} \simeq G_2$

$x = 10H_{\alpha_1} + 11H_{\alpha_2} + 16H_{\alpha_3} + 22H_{\alpha_4} + 17H_{\alpha_5} + 12H_{\alpha_6} + 3H_{\alpha_7}$

$e = \sqrt{5}(X_{\alpha_1+\alpha_2+\alpha_3+2\alpha_4+\alpha_5+\alpha_6+\alpha_7} + X_{\alpha_1+\alpha_2+2\alpha_3+2\alpha_4+2\alpha_5+\alpha_6+\alpha_7})+$
$\sqrt{8}(X_{-\alpha_6-\alpha_7} + X_{-\alpha_2-\alpha_3-2\alpha_4-2\alpha_5-\alpha_6-\alpha_7})$
$+3X_{\alpha_2+\alpha_3+2\alpha_4+2\alpha_5+2\alpha_6+\alpha_7}$

$\mathfrak{t}^1_{\mathbb{C}} = \mathbb{C}H_{\alpha_2} \oplus \mathbb{C}(H_{\alpha_3} + H_{\alpha_4})$

EVIII.

Let $\Delta = \{\alpha_1, \alpha_2, \ldots, \alpha_8\}$ be the Bourbaki simple roots of $\mathfrak{g}_{\mathbb{C}}$ then $\Delta_k = \{\beta_1, \ldots, \beta_8\}$, where $\beta_1 = 2\alpha_1 + 2\alpha_2 + 3\alpha_3 + 4\alpha_4 + 3\alpha_5 + 2\alpha_6 + \alpha_7$, $\beta_2 = \alpha_8$, $\beta_3 = \alpha_7$, $\beta_4 = \alpha_6$, $\beta_5 = \alpha_5$, $\beta_6 = \alpha_4$, $\beta_7 = \alpha_2$ and $\beta_8 = \alpha_3$, is a set of simple roots for $\mathfrak{k}_{\mathbb{C}} = \mathfrak{so}_{16}(\mathbb{C})$.

Table XI

---

4. 02000000 $\quad \mathfrak{k}_{\mathbb{C}}^{(x,e,f)} \simeq A_5 + A_1$

$x = 4H_{\alpha_1} + 6H_{\alpha_2} + 8H_{\alpha_3} + 12H_{\alpha_4} + 10H_{\alpha_5} + 8H_{\alpha_6} + 6H_{\alpha_7} + 4H_{\alpha_8}$

$e = \sqrt{2}(X_{\alpha_1+\alpha_2+2\alpha_3+3\alpha_4+2\alpha_5+2\alpha_6+2\alpha_7+\alpha_8} + X_{\alpha_1+2\alpha_2+2\alpha_3+3\alpha_4+3\alpha_5+2\alpha_6+\alpha_7+\alpha_8})$

$\mathfrak{t}^1_{\mathbb{C}} = \mathbb{C}H_{\alpha_1} \oplus \mathbb{C}H_{\alpha_3} \oplus \mathbb{C}H_{\alpha_6} \oplus \mathbb{C}(H_{\alpha_2} + H_{\alpha_4}) \oplus C(H_{\alpha_4} + H_{\alpha_5}) \oplus \mathbb{C}(H_{\alpha_5} + H_{\alpha_7})$

---

5. 00000020 $\quad \mathfrak{k}_{\mathbb{C}}^{(x,e,f)} \simeq C_4$

$x = 2H_{\alpha_1} + 6H_{\alpha_2} + 6H_{\alpha_3} + 10H_{\alpha_4} + 8H_{\alpha_5} + 6H_{\alpha_6} + 4H_{\alpha_7} + 2H_{\alpha_8}$

$e = X_{\alpha_1+2\alpha_2+2\alpha_3+3\alpha_4+2\alpha_5+\alpha_6} + X_{\alpha_1+2\alpha_2+2\alpha_3+3\alpha_4+2\alpha_5+2\alpha_6+2\alpha_7+\alpha_8}$
$+X_{\alpha_1+2\alpha_2+2\alpha_3+4\alpha_4+4\alpha_5+3\alpha_6+2\alpha_7+\alpha_8} + X_{-\alpha_1}$

$\mathfrak{t}^1_{\mathbb{C}} = \mathbb{C}H_{\alpha_4} \oplus \mathbb{C}H_{\alpha_6} \oplus \mathbb{C}H_{\alpha_8} \oplus \mathbb{C}(H_{\alpha_1} + H_{\alpha_2} + 2H_{\alpha_3} + 2H_{\alpha_5} + H_{\alpha_7})$

---

14. 40000000 $\quad \mathfrak{k}_{\mathbb{C}}^{(x,e,f)} \simeq 2G_2$

$x = 8H_{\alpha_1} + 10H_{\alpha_2} + 14H_{\alpha_3} + 20H_{\alpha_4} + 16H_{\alpha_5} + 12H_{\alpha_6} + 8H_{\alpha_7} + 4H_{\alpha_8}$

$e = \sqrt{2}(X_{\alpha_1} + X_{\alpha_1+\alpha_2+2\alpha_3+2\alpha_4+\alpha_5} + X_{\alpha_1+\alpha_2+2\alpha_3+3\alpha_4+3\alpha_5+3\alpha_6+2\alpha_7+\alpha_8}$
$+X_{\alpha_1+3\alpha_2+3\alpha_3+5\alpha_4+4\alpha_5+3\alpha_6+2\alpha_7+\alpha_8})$

$\mathfrak{t}^1_{\mathbb{C}} = \mathbb{C}H_{\alpha_4} \oplus \mathbb{C}H_{\alpha_5} \oplus \mathbb{C}H_{\alpha_7} \oplus \mathbb{C}H_{\alpha_8}$

---

15. 20000002 $\quad \mathfrak{k}_{\mathbb{C}}^{(x,e,f)} \simeq G_2$

$x = 6H_{\alpha_1} + 10H_{\alpha_2} + 14H_{\alpha_3} + 20H_{\alpha_4} + 16H_{\alpha_5} + 12H_{\alpha_6} + 8H_{\alpha_7} + 4H_{\alpha_8}$

$e = \sqrt{2}(X_{\alpha_1+\alpha_2+2\alpha_3+3\alpha_4+2\alpha_5+\alpha_6+\alpha_7+\alpha_8} + X_{\alpha_1+2\alpha_2+2\alpha_3+3\alpha_4+3\alpha_5+3\alpha_6+2\alpha_7+\alpha_8})$
$+X_{\alpha_1+2\alpha_2+2\alpha_3+4\alpha_4+3\alpha_5+2\alpha_6+\alpha_7} + X_{-\alpha_1} + X_{\alpha_1+\alpha_2+2\alpha_3+2\alpha_4+2\alpha_5+\alpha_6+\alpha_7}$
$+X_{\alpha_1+\alpha_2+2\alpha_3+2\alpha_4+\alpha_5+\alpha_6}$

$\mathfrak{t}^1_{\mathbb{C}} = \mathbb{C}(H_{\alpha_5} - H_{\alpha_7}) \oplus \mathbb{C}(H_{\alpha_2} - H_{\alpha_5} - H_{\alpha_6})$

---

16. 00020000 $\mathfrak{k}_{\mathbb{C}}^{(x,e,f)} \simeq 4A_1$

$x = 4H_{\alpha_1} + 8H_{\alpha_2} + 10H_{\alpha_3} + 16H_{\alpha_4} + 14H_{\alpha_5} + 12H_{\alpha_6} + 8H_{\alpha_7} + 4H_{\alpha_8}$

$e = \sqrt{2}(X_{\alpha_1+\alpha_2+\alpha_3+2\alpha_4+2\alpha_5+2\alpha_6+\alpha_7} + X_{\alpha_1+\alpha_2+2\alpha_3+2\alpha_4+2\alpha_5+2\alpha_6+\alpha_7+\alpha_8}$
$+X_{\alpha_1+2\alpha_2+2\alpha_3+4\alpha_4+3\alpha_5+2\alpha_6+2\alpha_7+\alpha_8} + X_{-\alpha_1})$

$\mathfrak{t}^1_{\mathbb{C}} = \mathbb{C}H_{\alpha_2} \oplus \mathbb{C}H_{\alpha_5} \oplus \mathbb{C}(H_{\alpha_4} - H_{\alpha_7}) \oplus \mathbb{C}(H_{\alpha_1} + 2H_{\alpha_3} + 3H_{\alpha_4} + H_{\alpha_6})$

---

19. 02000020 $\mathfrak{k}_{\mathbb{C}}^{(x,e,f)} \simeq B_2 + A_1$

$x = 6H_{\alpha_1} + 12H_{\alpha_2} + 14H_{\alpha_3} + 22H_{\alpha_4} + 18H_{\alpha_5} + 14H_{\alpha_6} + 10H_{\alpha_7} + 6H_{\alpha_8}$

$e = \sqrt{3}(X_{\alpha_1+2\alpha_2+2\alpha_3+4\alpha_4+3\alpha_5+2\alpha_6+\alpha_7} + X_{-\alpha_1-\alpha_3-\alpha_4-\alpha_5-\alpha_6-\alpha_7})$
$+2X_{\alpha_1+\alpha_2+2\alpha_3+2\alpha_4+2\alpha_5+2\alpha_6+2\alpha_7+\alpha_8} + X_{\alpha_1+\alpha_2+\alpha_3+2\alpha_4+\alpha_5+\alpha_6+\alpha_7+\alpha_8}$
$+X_{\alpha_1+\alpha_2+2\alpha_3+3\alpha_4+3\alpha_5+2\alpha_6+\alpha_7+\alpha_8}$

$\mathfrak{t}^1_{\mathbb{C}} = \mathbb{C}H_{\alpha_6} \oplus \mathbb{C}(H_{\alpha_3} + H_{\alpha_4}) \oplus \mathbb{C}(H_{\alpha_1} + H_{\alpha_2} + H_{\alpha_5})$

---

20. 00000200 $\mathfrak{k}_{\mathbb{C}}^{(x,e,f)} \simeq 4A_1$

$x = 4H_{\alpha_1} + 10H_{\alpha_2} + 12H_{\alpha_3} + 20H_{\alpha_4} + 16H_{\alpha_5} + 12H_{\alpha_6} + 8H_{\alpha_7} + 4H_{\alpha_8}$

$e = \sqrt{4-u}X_{\alpha_1+\alpha_2+2\alpha_3+3\alpha_4+2\alpha_5+\alpha_6} + 2X_{\alpha_1+2\alpha_2+2\alpha_3+3\alpha_4+2\alpha_5+2\alpha_6+2\alpha_7+\alpha_8}$
$+\sqrt{u}X_{\alpha_1+2\alpha_2+2\alpha_3+3\alpha_4+2\alpha_5+\alpha_6} + \sqrt{2-u}X_{-\alpha_1-\alpha_3-\alpha_4} + \sqrt{2+u}X_{-\alpha_1-\alpha_2-\alpha_3-\alpha_4}$

$\mathfrak{t}^1_{\mathbb{C}} = \mathbb{C}H_{\alpha_3} \oplus \mathbb{C}H_{\alpha_6} + \mathbb{C}H_{\alpha_8} \oplus \mathbb{C}(H_{\alpha_1} + H_{\alpha_2} + 2H_{\alpha_4} + 2H_{\alpha_5} + H_{\alpha_7})$

---

21. 02000040 $\mathfrak{k}_{\mathbb{C}}^{(x,e,f)} \simeq C_3 + A_1$

$x = 8H_{\alpha_1} + 18H_{\alpha_2} + 20H_{\alpha_3} + 32H_{\alpha_4} + 26H_{\alpha_5} + 20H_{\alpha_6} + 14H_{\alpha_7} + 8H_{\alpha_8}$

$e = \sqrt{6}(X_{\alpha_1+\alpha_2+\alpha_3+\alpha_4+\alpha_5+\alpha_6+\alpha_7+\alpha_8} + X_{\alpha_1+\alpha_2+2\alpha_3+3\alpha_4+2\alpha_5+\alpha_6+\alpha_7+\alpha_8}$
$+X_{\alpha_1+\alpha_2+2\alpha_3+3\alpha_4+3\alpha_5+3\alpha_6+2\alpha_7+\alpha_8}) + \sqrt{10}X_{-\alpha_1-\alpha_3-\alpha_4-\alpha_5-\alpha_6-\alpha_7-\alpha_8}$

$\mathfrak{t}^1_{\mathbb{C}} = \mathbb{C}H_{\alpha_3} \oplus \mathbb{C}H_{\alpha_5} + \mathbb{C}H_{\alpha_7} \oplus \mathbb{C}(H_{\alpha_1} + H_{\alpha_2} + 2H_{\alpha_4} + H_{\alpha_6})$

---

29. 02020000 $\mathfrak{k}_{\mathbb{C}}^{(x,e,f)} \simeq A_2 + A_1 + T_1$

$x = 8H_{\alpha_1} + 14H_{\alpha_2} + 18H_{\alpha_3} + 28H_{\alpha_4} + 24H_{\alpha_5} + 20H_{\alpha_6} + 14H_{\alpha_7} + 8H_{\alpha_8}$

$e = 2(X_{\alpha_1+\alpha_3+\alpha_4+\alpha_5+\alpha_6+\alpha_7+\alpha_8} + X_{\alpha_1+2\alpha_2+2\alpha_3+3\alpha_4+2\alpha_5+\alpha_6+\alpha_7+\alpha_8})$
$+\sqrt{6}(X_{\alpha_1+\alpha_2+2\alpha_3+3\alpha_4+3\alpha_5+2\alpha_6+\alpha_7} + X_{-\alpha_1-\alpha_3-\alpha_4-\alpha_5})$

$\mathfrak{t}^1_{\mathbb{C}} = \mathbb{C}H_{\alpha_3} \oplus \mathbb{C}H_{\alpha_4} \oplus \mathbb{C}H_{\alpha_7} \oplus \mathbb{C}(H_{\alpha_1} + H_{\alpha_2} + H_{\alpha_5} + H_{\alpha_6})$

---

30. 00020020 $\mathfrak{k}_{\mathbb{C}}^{(x,e,f)} \simeq B_2$

$x = 6H_{\alpha_1} + 14H_{\alpha_2} + 16H_{\alpha_3} + 26H_{\alpha_4} + 22H_{\alpha_5} + 18H_{\alpha_6} + 12H_{\alpha_7} + 6H_{\alpha_8}$

$e = 2(X_{\alpha_1+2\alpha_2+2\alpha_3+3\alpha_4+2\alpha_5+\alpha_6+\alpha_7+\alpha_8} + X_{-\alpha_1-\alpha_3-\alpha_4-\alpha_5-\alpha_6-\alpha_7-\alpha_8})$

$$+\sqrt{3}(X_{\alpha_1+\alpha_2+\alpha_3+2\alpha_4+2\alpha_5+2\alpha_6+\alpha_7}+X_{-\alpha_1-\alpha_2-\alpha_3-\alpha_4}$$
$$+X_{\alpha_1+\alpha_2+2\alpha_3+2\alpha_4+2\alpha_5+2\alpha_6+\alpha_7+\alpha_8}+X_{\alpha_1+\alpha_2+2\alpha_3+3\alpha_4+2\alpha_5+2\alpha_6+2\alpha_7+\alpha_8})$$

$$\mathfrak{t}^1_{\mathbb{C}}=\mathbb{C}(H_{\alpha_1}+H_{\alpha_2}+2H_{\alpha_3}+3H_{\alpha_4}+2H_{\alpha_5}+H_{\alpha_6})\oplus\mathbb{C}(H_{\alpha_4}-H_{\alpha_7})$$

---

34. 00000004 $\mathfrak{k}^{(x,e,f)}_{\mathbb{C}}\simeq A_2$

$$x=4H_{\alpha_1}+10H_{\alpha_2}+14H_{\alpha_3}+20H_{\alpha_4}+16H_{\alpha_5}+12H_{\alpha_6}+8H_{\alpha_7}+4H_{\alpha_8}$$

$$e=\sqrt{2-u}X_{\alpha_1+\alpha_2+2\alpha_3+2\alpha_4+\alpha_5}+\sqrt{2+u}X_{\alpha_1+\alpha_2+2\alpha_3+2\alpha_4+\alpha_5+\alpha_6}$$
$$+\sqrt{2}X_{-\alpha_1-\alpha_3-\alpha_4-\alpha_5-\alpha_6-\alpha_7}+\sqrt{4-u}X_{-\alpha_1-\alpha_2-\alpha_3-\alpha_4-\alpha_5}$$
$$\sqrt{2}X_{\alpha_1+\alpha_2+2\alpha_3+3\alpha_4+3\alpha_5+2\alpha_6+\alpha_7}+2X_{\alpha_1+2\alpha_2+2\alpha_3+3\alpha_4+3\alpha_5+2\alpha_6+2\alpha_7+\alpha_8}$$
$$+\sqrt{u}X_{-\alpha_1-\alpha_2-\alpha_3-\alpha_4-\alpha_5-\alpha_6}$$

$$\mathfrak{t}^1_{\mathbb{C}}=\mathbb{C}(2H_{\alpha_1}+4H_{\alpha_2}+3H_{\alpha_3}+7H_{\alpha_4}+4H_{\alpha_5}+3H_{\alpha_6}+2H_{\alpha_7})\oplus\mathbb{C}(H_{\alpha_2}+H_{\alpha_4}-H_{\alpha_8})$$

---

35. 200020002 $\mathfrak{k}^{(x,e,f)}_{\mathbb{C}}\simeq A_1+T_1$

$$x=8H_{\alpha_1}+14H_{\alpha_2}+18H_{\alpha_3}+28H_{\alpha_4}+24H_{\alpha_5}+18H_{\alpha_6}+12H_{\alpha_7}+6H_{\alpha_8}$$

$$e=\sqrt{2+u}X_{\alpha_1+\alpha_2+\alpha_3+2\alpha_4+2\alpha_5+\alpha_6}+\sqrt{2-u}X_{\alpha_1+\alpha_2+2\alpha_3+2\alpha_4+2\alpha_5+\alpha_6}$$
$$+2X_{\alpha_1+\alpha_2+2\alpha_3+2\alpha_4+2\alpha_5+2\alpha_6+2\alpha_7+\alpha_8}+\sqrt{4-u}X_{-\alpha_1-\alpha_3}$$
$$\sqrt{2}X_{\alpha_1+2\alpha_2+2\alpha_3+3\alpha_4+2\alpha_5+\alpha_6+\alpha_7}+\sqrt{2}X_{\alpha_1+\alpha_2+2\alpha_3+3\alpha_4+2\alpha_5+2\alpha_6+\alpha_7+\alpha_8}$$
$$+\sqrt{u}X_{-\alpha_1}$$

$$\mathfrak{t}^1_{\mathbb{C}}=\mathbb{C}(H_{\alpha_2}+H_{\alpha_6})\oplus\mathbb{C}(H_{\alpha_2}+H_{\alpha_8})$$

---

36. 00200002 $\mathfrak{k}^{(x,e,f)}_{\mathbb{C}}\simeq A_1$

$$x=6H_{\alpha_1}+12H_{\alpha_2}+16H_{\alpha_3}+24H_{\alpha_4}+20H_{\alpha_5}+16H_{\alpha_6}+12H_{\alpha_7}+6H_{\alpha_8}$$

$$e=X_{\alpha_1+\alpha_2+2\alpha_3+2\alpha_4+2\alpha_5+\alpha_6+\alpha_7}+X_{\alpha_1+\alpha_2+\alpha_3+2\alpha_4+2\alpha_5+2\alpha_6+2\alpha_7+\alpha_8}$$
$$+\sqrt{3}(X_{\alpha_1+\alpha_2+2\alpha_3+2\alpha_4+\alpha_5+\alpha_6+\alpha_7+\alpha_8}+X_{-\alpha_1-\alpha_2-\alpha_3-2\alpha_4-\alpha_5})$$
$$+\sqrt{2}(X_{\alpha_1+2\alpha_2+2\alpha_3+3\alpha_4+3\alpha_5+2\alpha_6+\alpha_7+\alpha_8}+X_{-\alpha_1-\alpha_2-\alpha_3-\alpha_4-\alpha_5-\alpha_6})$$
$$+2X_{\alpha_1+2\alpha_2+2\alpha_3+4\alpha_4+3\alpha_5+2\alpha_6+\alpha_7}$$

$$\mathfrak{t}^1_{\mathbb{C}}=\mathbb{C}(2H_{\alpha_1}+3H_{\alpha_3}+4H_{\alpha_4}+4H_{\alpha_5}+3H_{\alpha_6}+H_{\alpha_7}+H_{\alpha_8})$$

---

45. 00400000 $\mathfrak{k}^{(x,e,f)}_{\mathbb{C}}\simeq 2A_1$

$$x=8H_{\alpha_1}+14H_{\alpha_2}+18H_{\alpha_3}+28H_{\alpha_4}+24H_{\alpha_5}+20H_{\alpha_6}+16H_{\alpha_7}+8H_{\alpha_8}$$

$$e=2(X_{\alpha_1+\alpha_3+\alpha_4+\alpha_5+\alpha_6+\alpha_7}+X_{\alpha_1+2\alpha_2+2\alpha_3+3\alpha_4+2\alpha_5+\alpha_6+\alpha_7})$$
$$+\sqrt{2}(X_{\alpha_1+\alpha_2+\alpha_3+\alpha_4+\alpha_5+\alpha_6+\alpha_7+\alpha_8}+X_{-\alpha_1-\alpha_2-\alpha_3-\alpha_4-\alpha_5-\alpha_6})$$
$$+\sqrt{6}(X_{\alpha_1+2\alpha_2+3\alpha_3+4\alpha_4+3\alpha_5+2\alpha_6+\alpha_7+\alpha_8}+X_{-\alpha_1-\alpha_2-2\alpha_3-2\alpha_4-\alpha_5})$$

$$\mathfrak{t}^1_{\mathbb{C}}=\mathbb{C}H_{\alpha_5}\oplus\mathbb{C}(2H_{\alpha_1}+2H_{\alpha_2}+2H_{\alpha_3}+5H_{\alpha_4}+2H_{\alpha_6}+H_{\alpha_7}+H_{\alpha_8})$$

---

46. 02000200 $\mathfrak{k}^{(x,e,f)}_{\mathbb{C}}\simeq T_2$

$$x=8H_{\alpha_1}+16H_{\alpha_2}+20H_{\alpha_3}+32H_{\alpha_4}+26H_{\alpha_5}+20H_{\alpha_6}+14H_{\alpha_7}+8H_{\alpha_8}$$

$$e=2(X_{\alpha_1+\alpha_2+\alpha_3+2\alpha_4+\alpha_5+\alpha_6+\alpha_7+\alpha_8}+X_{\alpha_1+\alpha_2+2\alpha_3+2\alpha_4+2\alpha_5+2\alpha_6+\alpha_7+\alpha_8})$$
$$+\sqrt{2}(X_{\alpha_1+\alpha_2+2\alpha_3+3\alpha_4+2\alpha_5+\alpha_6}+X_{-\alpha_1-\alpha_3-\alpha_4-\alpha_5})$$
$$+\sqrt{6}(X_{\alpha_1+2\alpha_2+2\alpha_3+3\alpha_4+3\alpha_5+2\alpha_6+\alpha_7}+X_{-\alpha_1-\alpha_2-\alpha_3-\alpha_4-\alpha_5-\alpha_6})$$

$$\mathfrak{t}^1_{\mathbb{C}} = \mathbb{C}(H_{\alpha_1} + 2H_{\alpha_2} + 3H_{\alpha_3} + 2H_{\alpha_4} - H_{\alpha_6}) \oplus \mathbb{C}(2H_{\alpha_1} + 3H_{\alpha_2} + 5H_{\alpha_3} + 4H_{\alpha_4} + H_{\alpha_5} + H_{\alpha_7})$$

---

51. 40000040 $\quad \mathfrak{k}^{(x,e,f)}_{\mathbb{C}} \simeq A_2$

$$x = 12H_{\alpha_1} + 22H_{\alpha_2} + 26H_{\alpha_3} + 40H_{\alpha_4} + 32H_{\alpha_5} + 24H_{\alpha_6} + 16H_{\alpha_7} + 8H_{\alpha_8}$$

$$e = \sqrt{6}(X_{\alpha_1+\alpha_2+\alpha_3+\alpha_4} + X_{\alpha_1+\alpha_2+\alpha_3+2\alpha_4+2\alpha_5+2\alpha_6+\alpha_7} + X_{\alpha_1+\alpha_2+\alpha_3+2\alpha_4+2\alpha_5+\alpha_6+\alpha_7+\alpha_8}) + \sqrt{2}(X_{\alpha_1+\alpha_2+2\alpha_3+3\alpha_4+2\alpha_5+\alpha_6+\alpha_7} + X_{\alpha_1+\alpha_2+2\alpha_3+2\alpha_4+2\alpha_5+2\alpha_6+\alpha_7+\alpha_8}) + \sqrt{10}X_{-\alpha_1}$$

$$\mathfrak{t}^1_{\mathbb{C}} = \mathbb{C}(H_{\alpha_4} - H_{\alpha_7}) \oplus \mathbb{C}(H_{\alpha_6} + 2H_{\alpha_7} + H_{\alpha_8})$$

---

52. 00200022 $\quad \mathfrak{k}^{(x,e,f)}_{\mathbb{C}} \simeq A_1 + T_1$

$$x = 8H_{\alpha_1} + 18H_{\alpha_2} + 22H_{\alpha_3} + 34H_{\alpha_4} + 28H_{\alpha_5} + 22H_{\alpha_6} + 16H_{\alpha_7} + 8H_{\alpha_8}$$

$$e = \sqrt{6}(X_{\alpha_1+\alpha_2+2\alpha_3+2\alpha_4+\alpha_5+\alpha_6+\alpha_7} + X_{\alpha_1+\alpha_2+\alpha_3+2\alpha_4+2\alpha_5+2\alpha_6+2\alpha_7+\alpha_8} + X_{\alpha_1+\alpha_2+2\alpha_3+3\alpha_4+3\alpha_5+2\alpha_6+\alpha_7}) + \sqrt{2}(X_{\alpha_1+\alpha_2+2\alpha_3+3\alpha_4+2\alpha_5+2\alpha_6+\alpha_7+\alpha_8} + X_{-\alpha_1-\alpha_2-\alpha_3-2\alpha_4-\alpha_5-\alpha_6}) + \sqrt{10}X_{-\alpha_1-\alpha_3-\alpha_4-\alpha_5-\alpha_6-\alpha_7}$$

$$\mathfrak{t}^1_{\mathbb{C}} = \mathbb{C}(H_{\alpha_4} - H_{\alpha_6}) \oplus \mathbb{C}(2H_{\alpha_1} + 2H_{\alpha_2} + 3H_{\alpha_3} + 5H_{\alpha_4} + 3H_{\alpha_5} + H_{\alpha_7} + H_{\alpha_8})$$

---

53. 20002020 $\quad \mathfrak{k}^{(x,e,f)}_{\mathbb{C}} \simeq A_1$

$$x = 10H_{\alpha_1} + 20H_{\alpha_2} + 24H_{\alpha_3} + 38H_{\alpha_4} + 32H_{\alpha_5} + 24H_{\alpha_6} + 16H_{\alpha_7} + 8H_{\alpha_8}$$

$$e = \sqrt{7-u}(X_{\alpha_1+\alpha_2+2\alpha_3+3\alpha_4+2\alpha_5+\alpha_6} + \sqrt{6}X_{\alpha_1+\alpha_2+2\alpha_3+2\alpha_4+2\alpha_5+\alpha_6+\alpha_7} + \sqrt{7}X_{\alpha_1+\alpha_2+\alpha_3+2\alpha_4+2\alpha_5+\alpha_6+\alpha_7+\alpha_8}) + X_{\alpha_1+\alpha_2+2\alpha_3+2\alpha_4+2\alpha_5+2\alpha_6+\alpha_7+\alpha_8} + X_{-\alpha_1-\alpha_2-\alpha_3-\alpha_4} + \sqrt{12-u}X_{-\alpha_1-\alpha_3-\alpha_4-\alpha_5} + \sqrt{u}X_{\alpha_1+\alpha_2+2\alpha_3+3\alpha_4+2\alpha_5+2\alpha_6+\alpha_7} + \sqrt{u-2}X_{-\alpha_1-\alpha_3-\alpha_4-\alpha_5-\alpha_6-\alpha_7}$$

$$\mathfrak{t}^1_{\mathbb{C}} = \mathbb{C}(H_{\alpha_3} + H_{\alpha_4} + H_{\alpha_7} + H_{\alpha_8})$$

---

54. 02020020 $\quad \mathfrak{k}^{(x,e,f)}_{\mathbb{C}} \simeq 2A_1$

$$x = 10H_{\alpha_1} + 20H_{\alpha_2} + 24H_{\alpha_3} + 38H_{\alpha_4} + 32H_{\alpha_5} + 26H_{\alpha_6} + 18H_{\alpha_7} + 10H_{\alpha_8}$$

$$e = 3X_{\alpha_1+\alpha_2+\alpha_3+2\alpha_4+\alpha_5+\alpha_6+\alpha_7+\alpha_8} + \sqrt{5}X_{\alpha_1+2\alpha_2+2\alpha_3+3\alpha_4+2\alpha_5+\alpha_6} + X_{\alpha_1+\alpha_2+2\alpha_3+2\alpha_4+2\alpha_5+\alpha_6+\alpha_7+\alpha_8} + \sqrt{8}X_{\alpha_1+\alpha_2+2\alpha_3+3\alpha_4+3\alpha_5+2\alpha_6+\alpha_7} + \sqrt{5}X_{-\alpha_1-\alpha_3-\alpha_4-\alpha_5-\alpha_6} + \sqrt{8}X_{-\alpha_1-\alpha_2-\alpha_3-2\alpha_4-\alpha_5}$$

$$\mathfrak{t}^1_{\mathbb{C}} = \mathbb{C}(H_{\alpha_3} + H_{\alpha_4}) \oplus \mathbb{C}(H_{\alpha_1} + H_{\alpha_2} + H_{\alpha_5} + H_{\alpha_6} + H_{\alpha_7})$$

---

55. 00020200 $\quad \mathfrak{k}^{(x,e,f)}_{\mathbb{C}} \simeq 2A_1$

$$x = 8H_{\alpha_1} + 18H_{\alpha_2} + 22H_{\alpha_3} + 36H_{\alpha_4} + 30H_{\alpha_5} + 24H_{\alpha_6} + 16H_{\alpha_7} + 8H_{\alpha_8}$$

$$e = \sqrt{5}X_{\alpha_1+\alpha_2+2\alpha_3+3\alpha_4+2\alpha_5+\alpha_6} + \sqrt{8}X_{\alpha_1+\alpha_2+\alpha_3+2\alpha_4+2\alpha_5+2\alpha_6+\alpha_7+\alpha_8} + X_{\alpha_1+2\alpha_2+2\alpha_3+3\alpha_4+2\alpha_5+\alpha_6+\alpha_7+\alpha_8} + \sqrt{8}X_{\alpha_1+\alpha_2+2\alpha_3+2\alpha_4+2\alpha_5+2\alpha_6+2\alpha_7+\alpha_8} + \sqrt{5}X_{-\alpha_1-\alpha_2-\alpha_3-\alpha_4-\alpha_5-\alpha_6} + 3X_{-\alpha_1-\alpha_3-\alpha_4-\alpha_5-\alpha_6-\alpha_7-\alpha_8}$$

$$\mathfrak{t}^1_{\mathbb{C}} = \mathbb{C}(H_{\alpha_1} + H_{\alpha_2} + H_{\alpha_3} + 2H_{\alpha_4} + H_{\alpha_6} + H_{\alpha_7}) \oplus \mathbb{C}H_{\alpha_5}$$

---

56. 02020040 $\quad \mathfrak{k}^{(x,e,f)}_{\mathbb{C}} \simeq 3A_1$

$$x = 12H_{\alpha_1} + 26H_{\alpha_2} + 30H_{\alpha_3} + 48H_{\alpha_4} + 40H_{\alpha_5} + 32H_{\alpha_6} + 22H_{\alpha_7} + 12H_{\alpha_8}$$

$$e = \sqrt{8}X_{\alpha_1+\alpha_2+\alpha_3+2\alpha_4+\alpha_5+\alpha_6+\alpha_7+\alpha_8} + \sqrt{18}X_{\alpha_1+\alpha_2+2\alpha_3+2\alpha_4+2\alpha_5+\alpha_6+\alpha_7+\alpha_8} + \sqrt{10}X_{\alpha_1+\alpha_2+2\alpha_3+3\alpha_4+2\alpha_5+2\alpha_6+\alpha_7} + \sqrt{10}X_{-\alpha_1-\alpha_2-2\alpha_3-2\alpha_4-\alpha_5} \sqrt{14}X_{-\alpha_1-\alpha_3-\alpha_4-\alpha_5-\alpha_6-\alpha_7-\alpha_8}$$

$$\mathfrak{t}^1_{\mathbb{C}} = \mathbb{C}H_{\alpha_7} \oplus \mathbb{C}(H_{\alpha_4} + H_{\alpha_5}) \oplus \mathbb{C}(H_{\alpha_1} + H_{\alpha_2} + H_{\alpha_3} + H_{\alpha_6})$$

---

70. 40040000 $\mathfrak{k}^{(x,e,f)}_{\mathbb{C}} \simeq 2A_1$

$$x = 16H_{\alpha_1} + 26H_{\alpha_2} + 34H_{\alpha_3} + 52H_{\alpha_4} + 44H_{\alpha_5} + 36H_{\alpha_6} + 24H_{\alpha_7} + 12H_{\alpha_8}$$

$$e = \sqrt{12}X_{\alpha_1+\alpha_3+\alpha_4+\alpha_5+\alpha_6} + \sqrt{6}X_{\alpha_1+\alpha_2+\alpha_3+\alpha_4+\alpha_5+\alpha_6+\alpha_7} + \sqrt{6}X_{\alpha_1+\alpha_2+2\alpha_3+3\alpha_4+2\alpha_5+\alpha_6+\alpha_7} + \sqrt{12}X_{\alpha_1+2\alpha_2+2\alpha_3+3\alpha_4+2\alpha_5+\alpha_6+\alpha_7+\alpha_8} + \sqrt{10}(X_{-\alpha_1} + X_{-\alpha_1-\alpha_2-2\alpha_3-2\alpha_4-\alpha_5})$$

$$\mathfrak{t}^1_{\mathbb{C}} = \mathbb{C}H_{\alpha_5} \oplus \mathbb{C}(H_{\alpha_2} + H_{\alpha_4} - H_{\alpha_7} - H_{\alpha_8})$$

---

71. 02020200 $\mathfrak{k}^{(x,e,f)}_{\mathbb{C}} \simeq T_2$

$$x = 12H_{\alpha_1} + 24H_{\alpha_2} + 30H_{\alpha_3} + 48H_{\alpha_4} + 40H_{\alpha_5} + 32H_{\alpha_6} + 22H_{\alpha_7} + 12H_{\alpha_8}$$

$$e = \sqrt{12}X_{\alpha_1+\alpha_2+2\alpha_3+3\alpha_4+2\alpha_5+\alpha_6} + \sqrt{6}X_{\alpha_1+\alpha_2+2\alpha_3+2\alpha_4+\alpha_5+\alpha_6+\alpha_7+\alpha_8} + \sqrt{6}X_{\alpha_1+\alpha_2+\alpha_3+2\alpha_4+2\alpha_5+\alpha_6+\alpha_7+\alpha_8} + \sqrt{10}X_{\alpha_1+\alpha_2+2\alpha_3+2\alpha_4+2\alpha_5+2\alpha_6+\alpha_7} + \sqrt{10}X_{-\alpha_1-\alpha_2-2\alpha_3-2\alpha_4-\alpha_5} + \sqrt{12}X_{-\alpha_1-\alpha_3-\alpha_4-\alpha_5-\alpha_6}$$

$$\mathfrak{t}^1_{\mathbb{C}} = \mathbb{C}(H_{\alpha_2} - H_{\alpha_7}) \oplus \mathbb{C}(H_{\alpha_1} + H_{\alpha_2} + H_{\alpha_3} + 2H_{\alpha_4} + H_{\alpha_5} + H_{\alpha_6} + H_{\alpha_7})$$

---

77. 04020200 $\mathfrak{k}^{(x,e,f)}_{\mathbb{C}} \simeq A_1 + T_1$

$$x = 16H_{\alpha_1} + 30H_{\alpha_2} + 38H_{\alpha_3} + 60H_{\alpha_4} + 50H_{\alpha_5} + 40H_{\alpha_6} + 28H_{\alpha_7} + 16H_{\alpha_8}$$

$$e = 4X_{\alpha_1+\alpha_2+\alpha_3+\alpha_4+\alpha_5+\alpha_6+\alpha_7+\alpha_8} + \sqrt{22-u}X_{\alpha_1+\alpha_2+2\alpha_3+3\alpha_4+2\alpha_5+\alpha_6+\alpha_7} + \sqrt{12}X_{\alpha_1+\alpha_2+2\alpha_3+2\alpha_4+2\alpha_5+2\alpha_6+\alpha_7} + \sqrt{14-u}X_{-\alpha_1-\alpha_3-\alpha_4-\alpha_5-\alpha_6-\alpha_7} + \sqrt{12}X_{-\alpha_1-\alpha_2-2\alpha_3-2\alpha_4-\alpha_5} + \sqrt{8+u}X_{-\alpha_1-\alpha_2-\alpha_3-\alpha_4-\alpha_5-\alpha_6-\alpha_7} + \sqrt{u}X_{\alpha_1+2\alpha_2+2\alpha_3+3\alpha_4+2\alpha_5+\alpha_6+\alpha_7}$$

$$\mathfrak{t}^1_{\mathbb{C}} = \mathbb{C}(H_{\alpha_1} + H_{\alpha_2} + H_{\alpha_3} + 2H_{\alpha_4} + H_{\alpha_6}) \oplus \mathbb{C}H_{\alpha_5}$$

---

78. 02020220 $\mathfrak{k}^{(x,e,f)}_{\mathbb{C}} \simeq A_1$

$$x = 14H_{\alpha_1} + 30H_{\alpha_2} + 36H_{\alpha_3} + 58H_{\alpha_4} + 48H_{\alpha_5} + 38H_{\alpha_6} + 26H_{\alpha_7} + 14H_{\alpha_8}$$

$$e = \sqrt{7}X_{\alpha_1+\alpha_2+2\alpha_3+3\alpha_4+2\alpha_5+\alpha_6} + \sqrt{15}X_{\alpha_1+\alpha_2+2\alpha_3+2\alpha_4+\alpha_5+\alpha_6+\alpha_7+\alpha_8} + \sqrt{15}X_{\alpha_1+\alpha_2+\alpha_3+2\alpha_4+2\alpha_5+\alpha_6+\alpha_7+\alpha_8} + \sqrt{12}X_{\alpha_1+\alpha_2+\alpha_3+2\alpha_4+2\alpha_5+2\alpha_6+\alpha_7} + 4X_{-\alpha_1-\alpha_3-\alpha_4-\alpha_5-\alpha_6-\alpha_7-\alpha_8} + \sqrt{7}X_{-\alpha_1-\alpha_2-\alpha_3-\alpha_4-\alpha_5-\alpha_6} + \sqrt{12}X_{-\alpha_1-\alpha_2-\alpha_3-2\alpha_4-\alpha_5}$$

$$\mathfrak{t}^1_{\mathbb{C}} = \mathbb{C}(H_{\alpha_1} + H_{\alpha_2} + 2H_{\alpha_3} + 2H_{\alpha_4} + 2H_{\alpha_5} + H_{\alpha_6} + H_{\alpha_7})$$

---

79. 020020220 $\mathfrak{k}^{(x,e,f)}_{\mathbb{C}} \simeq T_1$

$$x = 12H_{\alpha_1} + 26H_{\alpha_2} + 32H_{\alpha_3} + 50H_{\alpha_4} + 42H_{\alpha_5} + 32H_{\alpha_6} + 22H_{\alpha_7} + 12H_{\alpha_8}$$

$$e = \sqrt{2}X_{\alpha_1+\alpha_2+2\alpha_3+2\alpha_4+2\alpha_5+\alpha_6} + \sqrt{8}X_{\alpha_1+\alpha_2+2\alpha_3+2\alpha_4+\alpha_5+\alpha_6+\alpha_7+\alpha_8}$$

$$+\sqrt{18}X_{\alpha_1+\alpha_2+\alpha_3+2\alpha_4+2\alpha_5+\alpha_6+\alpha_7+\alpha_8}+\sqrt{10}X_{\alpha_1+\alpha_2+2\alpha_3+3\alpha_4+2\alpha_5+2\alpha_6+\alpha_7}$$
$$+\sqrt{10}X_{-\alpha_1-\alpha_2-\alpha_3-2\alpha_4-\alpha_5}+\sqrt{2}X_{-\alpha_1-\alpha_2-\alpha_3-\alpha_4-\alpha_5-\alpha_6}$$
$$+\sqrt{14}X_{-\alpha_1-\alpha_3-\alpha_4-\alpha_5-\alpha_6-\alpha_7-\alpha_8}$$

$$\mathfrak{t}^1_{\mathbb{C}}=\mathbb{C}(2H_{\alpha_1}+2H_{\alpha_2}+3H_{\alpha_3}+3H_{\alpha_4}+3H_{\alpha_5}+2H_{\alpha_6}+3H_{\alpha_7})$$

---

80. 00400040 $\mathfrak{k}_{\mathbb{C}}^{(x,e,f)}\simeq T_1$

$$x=12H_{\alpha_1}+26H_{\alpha_2}+30H_{\alpha_3}+48H_{\alpha_4}+40H_{\alpha_5}+32H_{\alpha_6}+24H_{\alpha_7}+12H_{\alpha_8}$$

$$e=\sqrt{2}X_{\alpha_1+\alpha_2+\alpha_3+\alpha_4+\alpha_5+\alpha_6+\alpha_7}+\sqrt{8}X_{\alpha_1+\alpha_2+2\alpha_3+2\alpha_4+\alpha_5+\alpha_6+\alpha_7+\alpha_8}$$
$$+\sqrt{10}X_{\alpha_1+\alpha_2+2\alpha_3+3\alpha_4+2\alpha_5+2\alpha_6+\alpha_7}+\sqrt{18}X_{\alpha_1+\alpha_2+\alpha_3+2\alpha_4+2\alpha_5+\alpha_6+\alpha_7+\alpha_8}$$
$$+\sqrt{2}X_{-\alpha_1-\alpha_2-\alpha_3-\alpha_4-\alpha_5-\alpha_6}+\sqrt{10}X_{-\alpha_1-\alpha_2-\alpha_3-2\alpha_4-\alpha_5}$$
$$+\sqrt{14}X_{-\alpha_1-\alpha_3-\alpha_4-\alpha_5-\alpha_6-\alpha_7-\alpha_8}$$

$$\mathfrak{t}^1_{\mathbb{C}}=\mathbb{C}(2H_{\alpha_1}+2H_{\alpha_2}+5H_{\alpha_3}+5H_{\alpha_4}+5H_{\alpha_5}+2H_{\alpha_6}+H_{\alpha_7})$$

---

82. 04020240 $\mathfrak{k}_{\mathbb{C}}^{(x,e,f)}\simeq 2A_1$

$$x=20H_{\alpha_1}+42H_{\alpha_2}+50H_{\alpha_3}+80H_{\alpha_4}+66H_{\alpha_5}+52H_{\alpha_6}+36H_{\alpha_7}+20H_{\alpha_8}$$

$$e=\sqrt{30}X_{\alpha_1+\alpha_2+2\alpha_3+3\alpha_4+2\alpha_5+\alpha_6+\alpha_7}+\sqrt{42}X_{\alpha_1+\alpha_2+\alpha_3+\alpha_4+\alpha_5+\alpha_6+\alpha_7+\alpha_8}$$
$$+4(X_{\alpha_1+\alpha_2+2\alpha_3+2\alpha_4+2\alpha_5+2\alpha_6+\alpha_7}+X_{-\alpha_1-\alpha_2-2\alpha_3-2\alpha_4-\alpha_5})$$
$$+\sqrt{30}X_{-\alpha_1-\alpha_2-\alpha_3-\alpha_4-\alpha_5-\alpha_6-\alpha_7}+\sqrt{22}X_{-\alpha_1-\alpha_3-\alpha_4-\alpha_5-\alpha_6-\alpha_7-\alpha_8}$$

$$\mathfrak{t}^1_{\mathbb{C}}=\mathbb{C}(H_{\alpha_1}+H_{\alpha_2}+H_{\alpha_3}+2H_{\alpha_4}+H_{\alpha_6})\oplus\mathbb{C}H_{\alpha_5}$$

---

93. 02022022 $\mathfrak{k}_{\mathbb{C}}^{(x,e,f)}\simeq T_1$

$$x=16H_{\alpha_1}+34H_{\alpha_2}+42H_{\alpha_3}+66H_{\alpha_4}+56H_{\alpha_5}+44H_{\alpha_6}+30H_{\alpha_7}+16H_{\alpha_8}$$

$$e=\sqrt{2-u}X_{\alpha_1+\alpha_2+2\alpha_3+2\alpha_4+2\alpha_5+\alpha_6+\alpha_7}+\sqrt{24}X_{\alpha_1+\alpha_2+2\alpha_3+2\alpha_4+\alpha_5+\alpha_6+\alpha_7+\alpha_8}$$
$$+\sqrt{28}X_{\alpha_1+\alpha_2+\alpha_3+2\alpha_4+2\alpha_5+2\alpha_6+\alpha_7}+\sqrt{14+u}X_{-\alpha_1-\alpha_2-\alpha_3-2\alpha_4-\alpha_5-\alpha_6}$$
$$+\sqrt{16-u}X_{-\alpha_1-\alpha_2-\alpha_3-\alpha_4-\alpha_5-\alpha_6-\alpha_7}+\sqrt{18}X_{-\alpha_1-\alpha_3-\alpha_4-\alpha_5-\alpha_6-\alpha_7-\alpha_8}$$
$$+\sqrt{10}X_{\alpha_1+\alpha_2+\alpha_3+2\alpha_4+2\alpha_5+\alpha_6+\alpha_7+\alpha_8}+\sqrt{u}X_{\alpha_1+\alpha_2+2\alpha_3+3\alpha_4+2\alpha_5+\alpha_6}$$

$$\mathfrak{t}^1_{\mathbb{C}}=\mathbb{C}(H_{\alpha_4}+H_{\alpha_7})$$

---

94. 40040040 $\mathfrak{k}_{\mathbb{C}}^{(x,e,f)}\simeq T_1$

$$x=20H_{\alpha_1}+38H_{\alpha_2}+46H_{\alpha_3}+72H_{\alpha_4}+60H_{\alpha_5}+48H_{\alpha_6}+32H_{\alpha_7}+16H_{\alpha_8}$$

$$e=\sqrt{30+u}X_{\alpha_1+\alpha_2+2\alpha_3+2\alpha_4+2\alpha_5+\alpha_6}+\sqrt{24}X_{\alpha_1+\alpha_2+2\alpha_3+2\alpha_4+\alpha_5+\alpha_6+\alpha_7}$$
$$+\sqrt{10}X_{\alpha_1+\alpha_2+\alpha_3+2\alpha_4+2\alpha_5+\alpha_6+\alpha_7}+\sqrt{18}X_{-\alpha_1-\alpha_3-\alpha_4-\alpha_5-\alpha_6-\alpha_7}$$
$$+\sqrt{-14-u}X_{\alpha_1+\alpha_2+\alpha_3+\alpha_4+\alpha_5+\alpha_6}+\sqrt{28}X_{-\alpha_1-\alpha_2-2\alpha_3-2\alpha_4-\alpha_5}$$
$$+\sqrt{16+u}X_{\alpha_1+\alpha_2+\alpha_3+2\alpha_4+\alpha_5+\alpha_6+\alpha_7+\alpha_8}+\sqrt{u}X_{\alpha_1+\alpha_2+2\alpha_3+3\alpha_4+2\alpha_5+\alpha_6+\alpha_7+\alpha_8}$$

$$\mathfrak{t}^1_{\mathbb{C}}=\mathbb{C}(H_{\alpha_4}-H_{\alpha_7}-H_{\alpha_8})$$

EIX.

Let $\Delta = \{\alpha_1, \alpha_2, \dots, \alpha_8\}$ be the Bourbaki simple roots of $\mathfrak{g}_{\mathbb{C}}$ then $\Delta_k = \{\beta_1, \dots, \beta_8\}$, where $\beta_1 = \alpha_1$, $\beta_2 = \alpha_2$, $\beta_3 = \alpha_3$, $\beta_4 = \alpha_4$, $\beta_5 = \alpha_5$, $\beta_6 = \alpha_6$, $\beta_7 = \alpha_7$ and $\beta_8 = 2\alpha_1 + 3\alpha_2 + 4\alpha_3 + 6\alpha_4 + 5\alpha_5 + 4\alpha_6 + 3\alpha_7 + 2\alpha_8$, is a set of simple roots for $\mathfrak{k}_{\mathbb{C}} = E_7 \oplus \mathfrak{sl}_2(\mathbb{C})$.

Table XII

---

6. 0000000 4 $\quad \mathfrak{k}_{\mathbb{C}}^{(x,e,f)} \simeq E_6$

$$x = 4H_{\alpha_1} + 6H_{\alpha_2} + 8H_{\alpha_3} + 12H_{\alpha_4} + 10H_{\alpha_5} + 8H_{\alpha_6} + 6H_{\alpha_7} + 4H_{\alpha_8}$$

$$e = \sqrt{2}(X_{\alpha_8} + X_{2\alpha_1+3\alpha_2+4\alpha_3+6\alpha_4+5\alpha_5+4\alpha_6+3\alpha_7+\alpha_8})$$

$$\mathfrak{t}_{\mathbb{C}}^1 = \mathbb{C}H_{\alpha_1} \oplus \mathbb{C}H_{\alpha_2} \oplus \mathbb{C}H_{\alpha_3} \oplus \mathbb{C}H_{\alpha_4} \oplus \mathbb{C}H_{\alpha_5} \oplus \mathbb{C}H_{\alpha_6}$$

---

7. 0000002 2 $\quad \mathfrak{k}_{\mathbb{C}}^{(x,e,f)} \simeq F_4$

$$x = 4H_{\alpha_1} + 6H_{\alpha_2} + 8H_{\alpha_3} + 12H_{\alpha_4} + 10H_{\alpha_5} + 8H_{\alpha_6} + 6H_{\alpha_7} + 2H_{\alpha_8}$$

$$e = X_{\alpha_2+\alpha_3+2\alpha_4+2\alpha_5+2\alpha_6+2\alpha_7+\alpha_8} + X_{-\alpha_8} + X_{2\alpha_1+2\alpha_2+3\alpha_3+4\alpha_4+3\alpha_5+2\alpha_6+2\alpha_7+\alpha_8} + X_{2\alpha_1+3\alpha_2+4\alpha_3+6\alpha_4+5\alpha_5+4\alpha_6+2\alpha_7+\alpha_8}$$

$$\mathfrak{t}_{\mathbb{C}}^1 = \mathbb{C}H_{\alpha_2} \oplus \mathbb{C}H_{\alpha_3} \oplus \mathbb{C}H_{\alpha_4} \oplus \mathbb{C}H_{\alpha_5}$$

---

8. 2000000 0 $\quad \mathfrak{k}_{\mathbb{C}}^{(x,e,f)} \simeq E_6$

$$x = 4H_{\alpha_1} + 4H_{\alpha_2} + 6H_{\alpha_3} + 8H_{\alpha_4} + 6H_{\alpha_5} + 4H_{\alpha_6} + 2H_{\alpha_7}$$

$$e = \sqrt{2}(X_{-\alpha_8} + X_{2\alpha_1+2\alpha_2+3\alpha_3+4\alpha_4+3\alpha_5+2\alpha_6+\alpha_7+\alpha_8})$$

$$\mathfrak{t}_{\mathbb{C}}^1 = \mathbb{C}(H_{\alpha_1} + 2H_{\alpha_7} + H_{\alpha_8}) \oplus \mathbb{C}H_{\alpha_2} \oplus \mathbb{C}H_{\alpha_3} \oplus \mathbb{C}H_{\alpha_4} \oplus \mathbb{C}H_{\alpha_5} \oplus \mathbb{C}H_{\alpha_6}$$

---

14. 0000020 0 $\quad \mathfrak{k}_{\mathbb{C}}^{(x,e,f)} \simeq G_2 + 2A_1$

$$x = 4H_{\alpha_1} + 6H_{\alpha_2} + 8H_{\alpha_3} + 12H_{\alpha_4} + 10H_{\alpha_5} + 8H_{\alpha_6} + 4H_{\alpha_7}$$

$$e = \sqrt{2}(X_{\alpha_1+\alpha_2+2\alpha_3+3\alpha_4+3\alpha_5+3\alpha_6+2\alpha_7+\alpha_8} + X_{-\alpha_6-\alpha_7-\alpha_8} + X_{\alpha_1+3\alpha_2+3\alpha_3+5\alpha_4+4\alpha_5+3\alpha_6+2\alpha_7+\alpha_8} + X_{-\alpha_2-\alpha_3-2\alpha_4-2\alpha_5-\alpha_6-\alpha_7-\alpha_8})$$

$$\mathfrak{t}_{\mathbb{C}}^1 = \mathbb{C}H_{\alpha_3} \oplus \mathbb{C}H_{\alpha_4} \oplus \mathbb{C}H_{\alpha_7} \oplus \mathbb{C}(2H_{\alpha_1} + 2H_{\alpha_2} + 3H_{\alpha_5} + 2H_{\alpha_6} + H_{\alpha_8})$$

---

18. 0000020 4 $\quad \mathfrak{k}_{\mathbb{C}}^{(x,e,f)} \simeq D_4$

$$x = 8H_{\alpha_1} + 12H_{\alpha_2} + 16H_{\alpha_3} + 24H_{\alpha_4} + 20H_{\alpha_5} + 16H_{\alpha_6} + 10H_{\alpha_7} + 4H_{\alpha_8}$$

$$e = \sqrt{4-u}X_{\alpha_2+\alpha_3+2\alpha_4+2\alpha_5+2\alpha_6+\alpha_7+\alpha_8} + \sqrt{2-u}X_{-\alpha_8} \; 2X_{2\alpha_1+2\alpha_2+3\alpha_3+4\alpha_4+3\alpha_5+2\alpha_6+2\alpha_7+\alpha_8} + \sqrt{u}X_{\alpha_2+\alpha_3+2\alpha_4+2\alpha_5+2\alpha_6+2\alpha_7+\alpha_8} + \sqrt{2+u}X_{-\alpha_7-\alpha_8}$$

$$\mathfrak{t}_{\mathbb{C}}^1 = \mathbb{C}H_{\alpha_3} \oplus \mathbb{C}H_{\alpha_4} \oplus \mathbb{C}H_{\alpha_5} \oplus \mathbb{C}H_{\alpha_2}$$

---

19. 0000004 0 $\quad \mathfrak{k}_{\mathbb{C}}^{(x,e,f)} \simeq D_4$

$$x = 4H_{\alpha_1} + 6H_{\alpha_2} + 8H_{\alpha_3} + 12H_{\alpha_4} + 10H_{\alpha_5} + 8H_{\alpha_6} + 6H_{\alpha_7}$$

$$\begin{aligned} e = {} & 2X_{\alpha_2+\alpha_3+2\alpha_4+2\alpha_5+2\alpha_6+2\alpha_7+\alpha_8} + \sqrt{2+u}X_{-\alpha_2-\alpha_3-2\alpha_4-2\alpha_5-2\alpha_6-\alpha_7-\alpha_8} \\ & \sqrt{2-u}X_{2\alpha_1+3\alpha_2+4\alpha_3+6\alpha_4+5\alpha_5+4\alpha_6+2\alpha_7+\alpha_8} + \sqrt{u}X_{2\alpha_1+\alpha_2+\alpha_3+2\alpha_4+2\alpha_5 2\alpha_6+2\alpha_7+\alpha_8} \\ & +\sqrt{4-u}X_{-\alpha_7-\alpha_8} \end{aligned}$$

$$\mathfrak{t}_{\mathbb{C}}^1 = \mathbb{C}H_{\alpha_3} \oplus \mathbb{C}H_{\alpha_4} \oplus \mathbb{C}H_{\alpha_5} \oplus \mathbb{C}H_{\alpha_2}$$

---

20. 2000002 2 $\quad \mathfrak{k}_{\mathbb{C}}^{(x,e,f)} \simeq B_3$

$$x = 8H_{\alpha_1} + 10H_{\alpha_2} + 14H_{\alpha_3} + 20H_{\alpha_4} + 16H_{\alpha_5} + 12H_{\alpha_6} + 8H_{\alpha_7} + 2H_{\alpha_8}$$

$$\begin{aligned} e = {} & X_{2\alpha_1+2\alpha_2+3\alpha_3+4\alpha_4+3\alpha_5+2\alpha_6+\alpha_7+\alpha_8} + X_{-\alpha_5-\alpha_6-\alpha_7-\alpha_8} \\ & \sqrt{3}(X_{\alpha_1+2\alpha_2+3\alpha_3+4\alpha_4+3\alpha_5+2\alpha_6+2\alpha_7+\alpha_8} + X_{\alpha_1+2\alpha_2+2\alpha_3+4\alpha_4+3\alpha_5+3\alpha_6+2\alpha_7+\alpha_8}) \\ & +2X_{-\alpha_2-\alpha_3-2\alpha_4-\alpha_5-\alpha_6-\alpha_7-\alpha_8} \end{aligned}$$

$$\mathfrak{t}_{\mathbb{C}}^1 = \mathbb{C}H_{\alpha_2} \oplus \mathbb{C}(H_{\alpha_4} + H_{\alpha_5}) \oplus \mathbb{C}(H_{\alpha_3} + H_{\alpha_6})$$

---

21. 0000004 8 $\quad \mathfrak{k}_{\mathbb{C}}^{(x,e,f)} \simeq F_4$

$$x = 12H_{\alpha_1} + 18H_{\alpha_2} + 24H_{\alpha_3} + 36H_{\alpha_4} + 30H_{\alpha_5} + 24H_{\alpha_6} + 18H_{\alpha_7} + 8H_{\alpha_8}$$

$$\begin{aligned} e = {} & \sqrt{6}(X_{\alpha_7+\alpha_8} + X_{\alpha_2+\alpha_3+2\alpha_4+2\alpha_5+2\alpha_6+\alpha_7+\alpha_8} + X_{2\alpha_1+2\alpha_2+3\alpha_3+4\alpha_4+3\alpha_5+2\alpha_6+\alpha_7+\alpha_8}) \\ & +\sqrt{10}X_{-\alpha_8} \end{aligned}$$

$$\mathfrak{t}_{\mathbb{C}}^1 = \mathbb{C}H_{\alpha_2} \oplus \mathbb{C}H_{\alpha_3} \oplus \mathbb{C}H_{\alpha_4} \oplus \mathbb{C}H_{\alpha_5}$$

---

22. 2000004 4 $\quad \mathfrak{k}_{\mathbb{C}}^{(x,e,f)} \simeq B4$

$$x = 12H_{\alpha_1} + 16H_{\alpha_2} + 22H_{\alpha_3} + 32H_{\alpha_4} + 26H_{\alpha_5} + 20H_{\alpha_6} + 14H_{\alpha_7} + 4H_{\alpha_8}$$

$$\begin{aligned} e = {} & \sqrt{6}(X_{2\alpha_1+2\alpha_2+3\alpha_3+4\alpha_4+3\alpha_5+2\alpha_6+\alpha_7+\alpha_8} + X_{-\alpha_7-\alpha_8} \\ & +X_{-\alpha_2-\alpha_3-2\alpha_4-2\alpha_5-2\alpha_6-\alpha_7-\alpha_8}) \\ & +\sqrt{10}X_{\alpha_2+\alpha_3+2\alpha_4+2\alpha_5+2\alpha_6+2\alpha_7+\alpha_8} \end{aligned}$$

$$\mathfrak{t}_{\mathbb{C}}^1 = \mathbb{C}H_{\alpha_2} \oplus \mathbb{C}H_{\alpha_3} \oplus \mathbb{C}H_{\alpha_4} \oplus \mathbb{C}H_{\alpha_5}$$

---

24. 4000000 4 $\quad \mathfrak{k}_{\mathbb{C}}^{(x,e,f)} \simeq A_3$

$$x = 12H_{\alpha_1} + 14H_{\alpha_2} + 20H_{\alpha_3} + 28H_{\alpha_4} + 22H_{\alpha_5} + 16H_{\alpha_6} + 10H_{\alpha_7} + 4H_{\alpha_8}$$

$$\begin{aligned} e = {} & \sqrt{6}(X_{\alpha_1++\alpha_3+\alpha_4+\alpha_5+\alpha_6+\alpha_7+\alpha_8} + X_{\alpha_1+3\alpha_2+3\alpha_3+5\alpha_4+4\alpha_5+3\alpha_6+2\alpha_7+\alpha_8}) \\ & +2(X_{-\alpha_8} + X_{-\alpha_2-\alpha_3-2\alpha_4-2\alpha_5-2\alpha_6-2\alpha_7-\alpha_8}) \end{aligned}$$

$$\mathfrak{t}_{\mathbb{C}}^1 = \mathbb{C}H_{\alpha_3} \oplus \mathbb{C}H_{\alpha_4} \oplus \mathbb{C}H_{\alpha_5} \oplus \mathbb{C}H_{\alpha_6}$$

---

25. 2000020 0 $\quad \mathfrak{k}_{\mathbb{C}}^{(x,e,f)} \simeq A_3 + T_1$

$$x = 8H_{\alpha_1} + 10H_{\alpha_2} + 14H_{\alpha_3} + 20H_{\alpha_4} + 16H_{\alpha_5} + 12H_{\alpha_6} + 6H_{\alpha_7}$$

$$\begin{aligned} e = {} & \sqrt{6}(X_{\alpha_1+2\alpha_2+3\alpha_3+4\alpha_4+4\alpha_5+3\alpha_6+2\alpha_7+\alpha_8} + X_{-\alpha_1-\alpha_2-2\alpha_3-2\alpha_4-2\alpha_5-\alpha_6-\alpha_7-\alpha_8}) \\ & +2(X_{2\alpha_1+2\alpha_2+3\alpha_3+4\alpha_4+3\alpha_5+2\alpha_6+\alpha_7+\alpha_8} + X_{-\alpha_2-\alpha_3-2\alpha_4-2\alpha_5-2\alpha_6-\alpha_7-\alpha_8}) \end{aligned}$$

$$\mathfrak{t}^1_{\mathbb{C}} = \mathbb{C}H_{\alpha_2} \oplus \mathbb{C}(H_{\alpha_3}+H_{\alpha_4}) \oplus \mathbb{C}(H_{\alpha_1}+H_{\alpha_3}+2H_{\alpha_6}+2H_{\alpha_7}+H_{\alpha_8}) \oplus \mathbb{C}(H_{\alpha_4}+H_{\alpha_5})$$

---

28. 0002000 0 $\mathfrak{k}_{\mathbb{C}}^{(x,e,f)} \simeq 2A_1$

$$x = 8H_{\alpha_1} + 12H_{\alpha_2} + 16H_{\alpha_3} + 24H_{\alpha_4} + 18H_{\alpha_5} + 12H_{\alpha_6} + 6H_{\alpha_7}$$

$$\begin{aligned} e = {} & \sqrt{2}(X_{\alpha_1+2\alpha_2+3\alpha_3+4\alpha_4+3\alpha_5+2\alpha_6+\alpha_7+\alpha_8} + X_{-\alpha_1-\alpha_2-2\alpha_3-2\alpha_4-2\alpha_5-2\alpha_6-\alpha_7-\alpha_8}) \\ & +2(X_{2\alpha_1+2\alpha_2+3\alpha_3+4\alpha_4+3\alpha_5+2\alpha_6+2\alpha_7+\alpha_8} + X_{-\alpha_2-\alpha_3-2\alpha_4-2\alpha_5-2\alpha_6-2\alpha_7-\alpha_8}) \\ & +\sqrt{6}(X_{-\alpha_1-\alpha_2-\alpha_3-2\alpha_4-2\alpha_5-\alpha_6-\alpha_7-\alpha_8} + X_{\alpha_1+2\alpha_2+2\alpha_3+4\alpha_4+4\alpha_5+3\alpha_6+2\alpha_7+\alpha_8}) \end{aligned}$$

$$\mathfrak{t}^1_{\mathbb{C}} = \mathbb{C}H_{\alpha_2} \oplus \mathbb{C}(H_{\alpha_1} + H_{\alpha_3} + 3H_{\alpha_4} + H_{\alpha_5} + 2H_{\alpha_6} + H_{\alpha_7} + H_{\alpha_8})$$

---

30. 2000022 2 $\mathfrak{k}_{\mathbb{C}}^{(x,e,f)} \simeq G_2$

$$x = 12H_{\alpha_1} + 16H_{\alpha_2} + 22H_{\alpha_3} + 32H_{\alpha_4} + 26H_{\alpha_5} + 20H_{\alpha_6} + 12H_{\alpha_7} + 2H_{\alpha_8}$$

$$\begin{aligned} e = {} & \sqrt{5}X_{\alpha_1+\alpha_2+2\alpha_3+2\alpha_4+2\alpha_5+2\alpha_6+2\alpha_7+\alpha_8} + \sqrt{8}X_{-\alpha_1-\alpha_2-\alpha_3-2\alpha_4-2\alpha_5-\alpha_6-\alpha_7-\alpha_8} \\ & +\sqrt{5}X_{\alpha_1+2\alpha_2+2\alpha_3+4\alpha_4+3\alpha_5+2\alpha_6+2\alpha_7+\alpha_8} + \sqrt{8}X_{-\alpha_1-\alpha_2-2\alpha_3-2\alpha_4-\alpha_5-\alpha_6-\alpha_7-\alpha_8} \\ & +3X_{2\alpha_1+2\alpha_2+3\alpha_3+4\alpha_4+3\alpha_5+2\alpha_6+\alpha_7+\alpha_8} + X_{-\alpha_2-\alpha_3-2\alpha_4-2\alpha_5-2\alpha_6-\alpha_7-\alpha_8} \end{aligned}$$

$$\mathfrak{t}^1_{\mathbb{C}} = \mathbb{C}H_{\alpha_2} \oplus \mathbb{C}(H_{\alpha_3} + H_{\alpha_4} + H_{\alpha_5})$$

---

31. 0000040 4 $\mathfrak{k}_{\mathbb{C}}^{(x,e,f)} \simeq G_2$

$$x = 12H_{\alpha_1} + 18H_{\alpha_2} + 24H_{\alpha_3} + 36H_{\alpha_4} + 30H_{\alpha_5} + 24H_{\alpha_6} + 14H_{\alpha_7} + 4H_{\alpha_8}$$

$$\begin{aligned} e = {} & X_{\alpha_2+\alpha_3+2\alpha_4+2\alpha_5+2\alpha_6+\alpha_7+\alpha_8} + \sqrt{5}X_{\alpha_1+\alpha_2+2\alpha_3+2\alpha_4+2\alpha_5+2\alpha_6+2\alpha_7+\alpha_8} \\ & +\sqrt{8}(X_{-\alpha_1-2\alpha_2-2\alpha_3-3\alpha_4-2\alpha_5-\alpha_6-\alpha_7-\alpha_8} + X_{-\alpha_1-\alpha_3-\alpha_4-\alpha_5-\alpha_6-\alpha_7-\alpha_8}) \\ & +\sqrt{5}X_{\alpha_1+2\alpha_2+2\alpha_3+4\alpha_4+3\alpha_5+2\alpha_6+2\alpha_7+\alpha_8} + 3X_{2\alpha_1+2\alpha_2+3\alpha_3+4\alpha_4+3\alpha_5+2\alpha_6+\alpha_7+\alpha_8} \end{aligned}$$

$$\mathfrak{t}^1_{\mathbb{C}} = \mathbb{C}H_{\alpha_5} \oplus \mathbb{C}(H_{\alpha_3} + H_{\alpha_4})$$

---

32. 2000024 4 $\mathfrak{k}_{\mathbb{C}}^{(x,e,f)} \simeq B_3$

$$x = 16H_{\alpha_1} + 22H_{\alpha_2} + 30H_{\alpha_3} + 44H_{\alpha_4} + 36H_{\alpha_5} + 28H_{\alpha_6} + 18H_{\alpha_7} + 4H_{\alpha_8}$$

$$\begin{aligned} e = {} & \sqrt{14}X_{\alpha_2+\alpha_3+2\alpha_4+2\alpha_5+2\alpha_6+2\alpha_7+\alpha_8} + \sqrt{18}X_{2\alpha_1+2\alpha_2+3\alpha_3+4\alpha_4+3\alpha_5+2\alpha_6+\alpha_7+\alpha_8} \\ & +\sqrt{10}(X_{-\alpha_1-\alpha_2-2\alpha_3-2\alpha_4-\alpha_5-\alpha_6-\alpha_7-\alpha_8} + X_{-\alpha_1-\alpha_2-\alpha_3-2\alpha_4-2\alpha_5-\alpha_6-\alpha_7-\alpha_8}) \\ & +\sqrt{8}X_{-\alpha_2-\alpha_3-2\alpha_4-2\alpha_5-2\alpha_6-\alpha_7-\alpha_8} \end{aligned}$$

$$\mathfrak{t}^1_{\mathbb{C}} = \mathbb{C}H_{\alpha_2} \oplus \mathbb{C}H_{\alpha_4} \oplus \mathbb{C}(H_{\alpha_3} + H_{\alpha_5})$$

---

33. 4000004 8 $\mathfrak{k}_{\mathbb{C}}^{(x,e,f)} \simeq B_3$

$$x = 20H_{\alpha_1} + 26H_{\alpha_2} + 36H_{\alpha_3} + 52H_{\alpha_4} + 42H_{\alpha_5} + 32H_{\alpha_6} + 22H_{\alpha_7} + 8H_{\alpha_8}$$

$$\begin{aligned} e = {} & \sqrt{14}X_{\alpha_2+\alpha_3+2\alpha_4+2\alpha_5+2\alpha_6+2\alpha_7+\alpha_8} + \sqrt{10}X_{\alpha_1+\alpha_3+\alpha_4+\alpha_5+\alpha_6+\alpha_7+\alpha_8} \\ & +\sqrt{10}X_{\alpha_1+2\alpha_2+3\alpha_3+4\alpha_4+3\alpha_5+2\alpha_6+\alpha_7+\alpha_8} + \sqrt{18}X_{-\alpha_3-\alpha_4-\alpha_5-\alpha_6-\alpha_7-\alpha_8} \\ & +\sqrt{8}X_{-\alpha_2-\alpha_4-\alpha_5-\alpha_6-\alpha_7-\alpha_8} \end{aligned}$$

$$\mathfrak{t}^1_{\mathbb{C}} = \mathbb{C}H_{\alpha_4} \oplus \mathbb{C}H_{\alpha_5} \oplus \mathbb{C}H_{\alpha_6}$$

---

34. 0002020 0 $\mathfrak{k}_{\mathbb{C}}^{(x,e,f)} \simeq 2A_1$

$$x = 12H_{\alpha_1} + 18H_{\alpha_2} + 24H_{\alpha_3} + 36H_{\alpha_4} + 28H_{\alpha_5} + 20H_{\alpha_6} + 10H_{\alpha_7}$$

$$e = \sqrt{6}X_{\alpha_1+2\alpha_2+3\alpha_3+4\alpha_4+3\alpha_5+2\alpha_6+2\alpha_7+\alpha_8}+\sqrt{10}X_{\alpha_1+\alpha_2+2\alpha_3+3\alpha_4+3\alpha_5+3\alpha_6+2\alpha_7+\alpha_8}$$
$$+\sqrt{12}X_{\alpha_1+2\alpha_2+3\alpha_3+4\alpha_4+3\alpha_5+2\alpha_6+\alpha_7+\alpha_8}+\sqrt{12}X_{-\alpha_2-\alpha_3-2\alpha_4-2\alpha_5-2\alpha_6-\alpha_7-\alpha_8}$$
$$+\sqrt{10}X_{-\alpha_1-\alpha_2-2\alpha_3-3\alpha_4-2\alpha_5-\alpha_6-\alpha_7-\alpha_8}+\sqrt{6}X_{-\alpha_1-\alpha_2-\alpha_3-2\alpha_4-2\alpha_5-2\alpha_6-2\alpha_7-\alpha_8}$$

$$\mathfrak{t}_{\mathbb{C}}^1 = \mathbb{C}H_{\alpha_5} \oplus \mathbb{C}(2H_{\alpha_1} + 2H_{\alpha_2} + 3H_{\alpha_3} + 3H_{\alpha_4} + 2H_{\alpha_6} + H_{\alpha_7} + H_{\alpha_8})$$

---

35. 4000040 4 $\mathfrak{k}_{\mathbb{C}}^{(x,e,f)} \simeq A_2$

$$x = 20H_{\alpha_1} + 26H_{\alpha_2} + 36H_{\alpha_3} + 52H_{\alpha_4} + 42H_{\alpha_5} + 32H_{\alpha_6} + 18H_{\alpha_7} + 4H_{\alpha_8}$$

$$e = \sqrt{22}X_{\alpha_1+2\alpha_2+3\alpha_3+4\alpha_4+3\alpha_5+2\alpha_6+2\alpha_7+\alpha_8}+\sqrt{u}X_{\alpha_1+\alpha_2+\alpha_3+2\alpha_4+2\alpha_5+2\alpha_6+2\alpha_7+\alpha_8}$$
$$+\sqrt{22-u}X_{\alpha_1+\alpha_2+\alpha_3+2\alpha_4+2\alpha_5+2\alpha_6+\alpha_7+\alpha_8}$$
$$+\sqrt{12}X_{-\alpha_1-\alpha_2-\alpha_3-2\alpha_4-2\alpha_5-\alpha_6-\alpha_7-\alpha_8}+\sqrt{12}X_{-\alpha_1-\alpha_2-2\alpha_3-2\alpha_4-\alpha_5-\alpha_6-\alpha_7-\alpha_8}$$
$$+\sqrt{8-u}X_{-\alpha_2-\alpha_3-2\alpha_4-2\alpha_5-2\alpha_6-\alpha_7-\alpha_8}+\sqrt{8+u}X_{-\alpha_2-\alpha_3-2\alpha_4-2\alpha_5-2\alpha_6-2\alpha_7-\alpha_8}$$

$$\mathfrak{t}_{\mathbb{C}}^1 = \mathbb{C}H_{\alpha_2} \oplus \mathbb{C}H_{\alpha_4}$$

---

36. 4000044 8 $\mathfrak{k}_{\mathbb{C}}^{(x,e,f)} \simeq G_2$

$$x = 28H_{\alpha_1} + 38H_{\alpha_2} + 52H_{\alpha_3} + 76H_{\alpha_4} + 62H_{\alpha_5} + 48H_{\alpha_6} + 30H_{\alpha_7} + 8H_{\alpha_8}$$

$$e = \sqrt{30}X_{\alpha_1+\alpha_2+\alpha_3+2\alpha_4+2\alpha_5+2\alpha_6+\alpha_7+\alpha_8} + \sqrt{22}X_{\alpha_2+\alpha_3+2\alpha_4+2\alpha_5+2\alpha_6+2\alpha_7+\alpha_8}$$
$$+\sqrt{30}X_{\alpha_1+2\alpha_2+3\alpha_3+4\alpha_4+3\alpha_5+2\alpha_6+\alpha_7+\alpha_8} + 4X_{-\alpha_1-\alpha_3-\alpha_4-\alpha_5-\alpha_6-\alpha_7-\alpha_8}$$
$$+\sqrt{42}X_{-\alpha_2-\alpha_3-2\alpha_4-2\alpha_5-2\alpha_6-\alpha_7-\alpha_8} + 4X_{-\alpha_1-2\alpha_2-2\alpha_3-3\alpha_4-2\alpha_5-\alpha_6-\alpha_7-\alpha_8}$$

$$\mathfrak{t}_{\mathbb{C}}^1 = \mathbb{C}H_{\alpha_4} \oplus \mathbb{C}H_{\alpha_5}$$

References

[A-H-V] J. Adams. J.-S. Huang, D. A. Vogan,, *Functios on the Model Orbit in E_8*, AMS Journal of representation theory **2** (1998), 224-263.

[A-K] L. Auslander and B. Kostant, *Polarization and unitary representations of solvable Lie groups*, Invent. Math. **14** (1971), 255-354.

[B0] N. Bourbaki, *Groupes et Algèbre de Lie Chapitres 4,5,6*, Elements de mathématique. **MASSON** (1981).

[C] C. W. Curtis, *Corrections and additions to: 'On the degrees and rationality of certain characters of finite Chevalley groups'*, Trans. Amer. Math. Soc. **202** (1975), 405-406.

[D] M. Duflo, *Construction de représentations unitaires d'un groupe de Lie*, Harmonic Analysis and Group Representations. **C.I.M.E.** (1982).

[D1] D. Djoković, *Classification of nilpotent elements in simple exceptional real Lie algebras of inner type and description of their centralizers*, J. Alg. **112** (1987), 577-585.

[D2] D. Djoković, *Classification of nilpotent elements in simple real Lie algebras $E_{6(6)}$ and $E_{6(-26)}$ and description of their centralizers*, J. Alg. **116** (1988), 196-207.

[D3] D. Djoković, *Proof of a conjecture of Kostant*, Trans. AMS **302(2)** (1988), 503-524.

[Dy] E. Dynkin, *Semisimple subalgebras of simple Lie algebras*, Selected papers of E.B. Dynkin with commentary edited by Yushkevich, Seitz and Onishchik AMS, (2000), 175-309.

[Dy] E. Dynkin, *Semisimple subalgebras of simple Lie algebras*, Amer. Soc. Transl. Ser. 2 **6,** (1957), 111-245.

[K] A. A. Kirillov, *Unitary representations of nilpotent Lie groups*, Russian Math. Surveys **17** (1962), 57-110.

[Kn] A. W. Knapp, *Lie Groups Beyond an Introduction*, vol. **140**, Birkhäuser, Progress in Mathematics Boston, 1996.

[K1] D. R. King, *The Component Groups of Nilpotents in Exceptional Simple Real Lie Algebras* , Communications In Algebra **20 (1)** (1992), 219-284.

[K2] D. R. King, *Spherical nilpotent orbits and the Kostant-Sekiguchi correspondence*, TAMS **34 (12)** (2002), 4909-4920.

[K3] D. R. King, *Classification of spherical nilpotent orbits in complex symmetric space*, J. Lie Theory **14** (2004), 339-370.

[K-R] B. Kostant, S. Rallis, *Orbits and Representations associated with symmetric spaces*, Amer. J. Math. **93** (1971), 753-809.

[McG] W. McGovern, *Ring of regular fuctions on nilpotent orbits II: Model Algebras and Orbits* , Comm. Alg., **22(3),** (1994), 765-772.

[No] A. G. Noel, *Nilpotent orbits and theta-stable parabolic Subalgebras*, AMS Journal of representation theory **2,** (1998), 1-32.

[No1] A. G. Noël, *Classification of Admissible Nilpotent Orbits In simple Exceptional real Lie algebras of Inner type*, AMS Journal of Representation Theory **5** (2001), 455-493.

[No2] A. G. Noël, *Classification of Admissible Nilpotent Orbits In simple real Lie algebras* $E_{6(6)}$ *and* $E_{6(-26)}$, AMS Journal of Representation Theory **5** (2001), 494-502.

[No3] A. G. Noël, *Computing Maximal Tori Using LiE and Mathematica* , Lecture Notes in Computer Science **2657** (2003), Springer-Verlag Berlin Heidelberg 2003, 728-736.

[Pa] D. I. Panyushev, *Complexity and nilpotent orbits*, Manuscripta Math. **83,** (1994), 223-237.

[O] T. Ohta, *Classification of admissible nilpotent orbits in the classical real Lie algebras*, J. of Algebra **136, No. 1** (1991), 290-333.

[Sch] J. Schwartz, *The determination of the admissible nilpotent orbits in real classical groups*, Ph. D. Thesis M.I.T. Cambridge, MA (1987).

[Se] J. Sekiguchi, *Remarks on real nilpotent orbits of a symmetric pair*, J. Math. Soc. Japan **39, No. 1** (1987), 127-138.

[S-S] T. A. Springer and R. Steinberg, *Seminar on Algebraic Groups and Related Finite Groups*, Lecture Notes in Mathematics **131** (1970), Springer-Verlag, Heidelberg.

[V]] D. Vogan jr, *Unitary representations of reductive groups*, Annals of Mathematical Studies, Princeton University Press Study **118** (1987).

[V1] D. Vogan jr, *Associated varieties and unipotent representations* , Harmonic Analysis on Reductive Groups (1991), Birkhäuser, Boston-Basel-Berlin, 315-388.

[V2] D. Vogan jr, *Singular unitary representations*, Non-commutative Harmonic Analysis and Lie groups, J. Carmona and M. Vergne, eds,Lecture Notes in Mathematics 880 (1987), Springer Verlag, Berlin-Heidelberg-New York, 506-535.

[VL] Van Leeuwen M. A. A., Cohen A. M., Lisser B., *LiE Apackage for Lie Group Computations*, Computer Algebra Nederland, Amsterdam The Netherlands (1992).

[W] Wolfram S., *The Mathematica Book*, Wolfram media, Cambridge University Press (1998).

MATHEMATICS DEPARTMENT THE UNIVERSITY OF MASSACHUSETTS, BOSTON, MA 02125
E-mail address: anoel@math.umb.edu

Contemporary Mathematics
Volume **467**, 2008

A Survey of CAARMS12 Participants

William A. Massey, Derrick Raphael, and Erica N. Walker

ABSTRACT. This article describes an exploratory study of the formative, school, and professional mathematics experiences of African American participants attending the twelfth Conference for African American Researchers in the Mathematical Sciences (CAARMS12). We report information about participants' parental background, secondary school experiences, participation in mathematics-based extracurricular activities, and college and graduate school awards in the mathematical sciences. While we do not claim these results to be representative of the experiences of all African Americans in the mathematical sciences, we believe that these preliminary findings are illuminating and provide directions for future research.

1. Introduction

Much of the research about African Americans in mathematics describes their under-performance in school mathematics. However, much can be learned from examining the experiences of those African American students who are successful in mathematics (Walker [**W6**], 2006). In particular, it may be useful to examine the formative, school, and professional experiences of African Americans who have successfully pursued undergraduate and graduate degrees in mathematics (Walker [**W8**], 2008), as there is evidence that compared to students from other ethnicities, they are less likely to persist in these programs (Herzig [**He**], 2004). While there are a number of college programs (for example, Treisman's Berkeley Calculus Workshops and the Meyerhoff Scholars Program at the University of Maryland, Baltimore County) that have been effective at improving outcomes in science, technology, engineering, and mathematics (STEM) fields for African Americans, there are also a substantial number of African American scholars who are successful in mathematics without participating in such programs. What experiences do successful African American scholars in the mathematical sciences report as formative to their pursuit of graduate degrees in the mathematical sciences and their professional development?

Here we report findings from an exploratory study. This is a survey of 45 attendees of the 12th Conference for African American Researchers in the Mathematical Sciences (CAARMS12), roughly half of the total number of conference attendees. There were 17 female and 28 male survey participants. This event was held June 20-23, 2006 at the University of North Carolina, Chapel Hill and

1991 *Mathematics Subject Classification.* Primary 01A80.
Key words and phrases. African-American, Diversity, Mathematicians, Survey.

SAMSI. This conference is also discussed in Massey [**M**]. The mission and impact of CAARMS conferences are discussed in the closing appendix for this paper. The goal of the study was to gain insight about the family, educational, and professional backgrounds and experiences of CAARMS participants.

Although this research is based on a snapshot of the conference attendees, it helps to shed much needed light on the characteristics of African Americans who have or plan to obtain a Ph.D. in the mathematical sciences. These conference attendees were given brief questionnaires to complete. They also participated in thirty to forty-five minute interviews, which enabled the interviewers to obtain an in-depth view of the lives of conference attendees. The following sections of this paper help to describe similarities and differences among this group of participants at the conference. We do not propose that sample used for the findings from this paper necessarily exhibit the characteristics of the population sought to study through any randomization process.

2. The survey questions and the responses.

Below are the survey questions:

(1) Where were attendees born and raised?
(2) How did individuals identify themselves?
(3) Where were participants' parents born and raised?
(4) What jobs did your parents have?
(5) What type of high schools did attendees go to?
(6) What high school interests did participants have?
(7) Participant involvement in various high school programs?
(8) What type of colleges did attendees go to?
(9) What were participants' undergraduate majors?
(10) What awards did participants win for college and graduate work?
(11) Which degrees did participants already have?
(12) Where will participants pursue a Ph.D.?

The following subsections discuss the responses to each question.

2.1. Where were attendees born and raised? In addition to the high numbers of African American attendees hailing from the United States of America (about 82%), a significant proportion of those in attendance were also directly from Africa (13%), see Figure 1. The important note to mention about those numbers is that the gender breakdown for African American attendees was about equal, while those from Africa were skewed in the direction of primarily male attendees. These characteristics distinguishing African Americans and Africans in the mathematical sciences also exist when all the numbers of these two groups are tallied nationwide as shown in Figures 13 and 14 of the appendix.

2.2. How did individuals identify themselves? Once again the number of participants who identified as African American was higher than any other group with about 82% selecting that category, see Figure 2. It is interesting however to also see that the percentage of those born and raised in Africa dropped to 9% from 13% while the percentage for Caribbean heritage increased. Thus, it appears that respondents who were born in Africa may also have parents with Caribbean heritage.

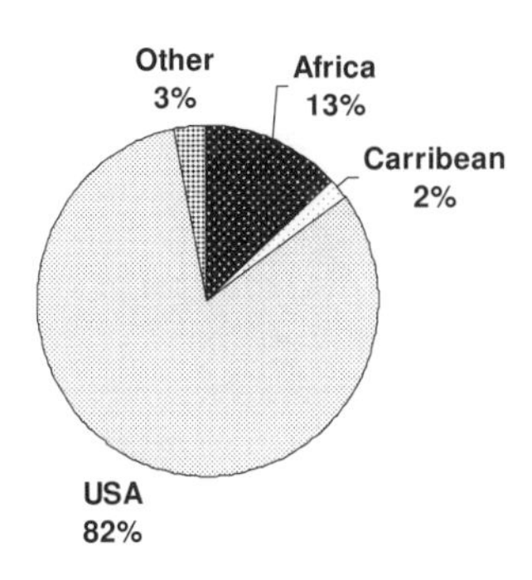

FIGURE 1. Survey question about the origins of the CAARMS12 attendees.

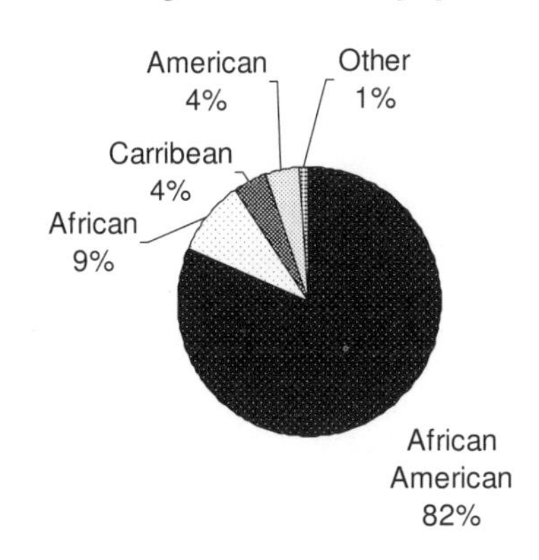

FIGURE 2. The self identities of the CAARMS12 attendees.

2.3. Where were participants' parents born and raised? Seventy three percent of attendees reported that their parents were born and raised in the United States, with 13% reporting that their parents were born in Africa, and 7% reporting parents who were born in the Caribbean, see Figure 3. Seven percent of the respondents also reported that their parents were from some other country. It is unclear from the data how many participants belong to households where both parents were from the same country, or two different countries. However, these results in this sections and the previous one reveal that this group of "African American" attendees is a diverse group.

2.4. What jobs did your parents have? There is one major characteristic that was brought to light through the analysis of the jobs held by the parents of conference participants. This was the relatively middle to upper middle class status held by these individuals. About 56% of participants' mothers were educators or held corporate office positions, while about 60% of the fathers of participants in the study were educators or government officials, see the graphs in Figure 4. These

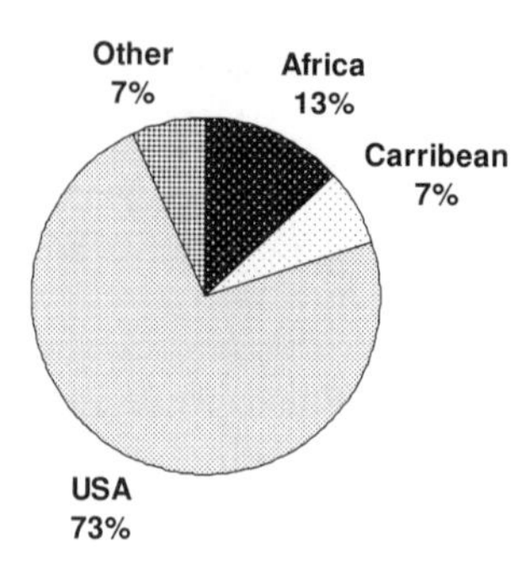

FIGURE 3. The origins of the parents of the CAARMS12 attendees.

high numbers of well educated and relatively affluent parents help to show why these individuals sought to obtain a Ph.D.

Thus, participants' parents were well educated and middle to upper middle class. Many researchers suggest that for African American students, a key determinant of success in school and college is the presence of networks that support academic engagement. Well educated and affluent parents may have been better able to expose their children to key opportunities that supported their mathematics development. These supplementary educational activities (Gordon, Bridglall, & Meroe [**G**], 2005) outside of school are critical to high academic achievement for all children, but are especially critical for students from underserved groups.

2.5. What type of high schools did attendees go to? When looking over the data, it was quite revealing that most of the CAARMS participants attended public schools, see Figure 5. Of the conference attendees, 80% went to public schools. One reason for this high percentage could be due to the more nurturing environment that may have been fostered at the public schools they attended, which could have differed from a more competitive atmosphere at a private institution. Information gleaned from the in-depth interviews revealed that attending private school instead of a public one may not be in the best interest for everyone, as one individual recounted the increased academic hardship encountered at such schools. This participant, while at public school, had a perfect academic record which was lowered substantially when making the transition to private school. Now 7% of attendees attended a religious school, which may have been either a public or private school. These institutions had an important impact on the attendees as these institutions impart a particular emphasis on organization and discipline that may not have been fostered as heavily at a purely public or private school. Overall the schools that individuals attended left an important imprint on their lives and future developments.

2.6. What high school interests did participants have? Surprisingly the most popular activity for the conference attendees was not involvement in math or science based activities but sports, see Figure 6. Over 60% of the participants were

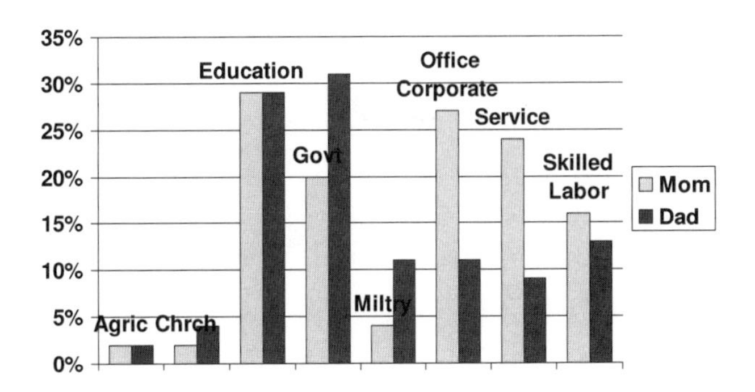

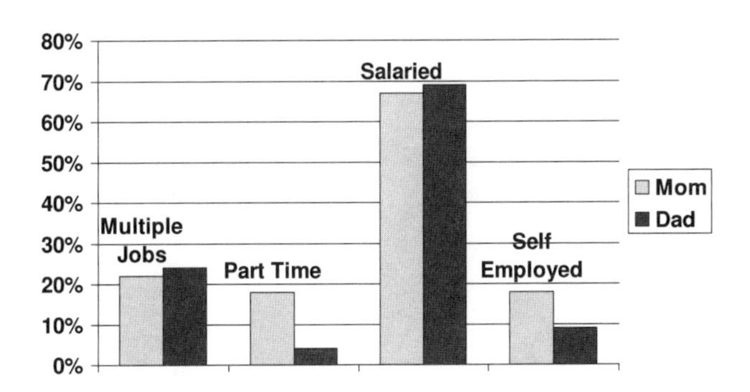

FIGURE 4. Survey questions about the jobs of the parents of the CAARMS12 attendees.

involved in some sports activity. The second most popular activity was the National Honor Society (52%) which shows the importance of early identification for academic excellence. Third was a math/science based club (47%). The fourth most popular high school activity was a tie between (music, band or chorus) and community service at 40% showing a connection between achieving personal excellence and giving back to others.

2.7. Participant involvement in various high school programs. Many conference participants showed an early interest in math or science based groups in high school, see Figure 7. However, only about 38% actually participated in a math-based summer program. Moreover, a mere 24% took part in an SAT prep program. This may be showing the relative strength of these individuals academically. Furthermore, they may not need the advantages to be garnered through participation in such activities. In addition, it may also be that students were not aware of math or science programs in high school, or that they chose not to participate. Further study of these issues would illuminate reasons for participation or lack there of in such programs.

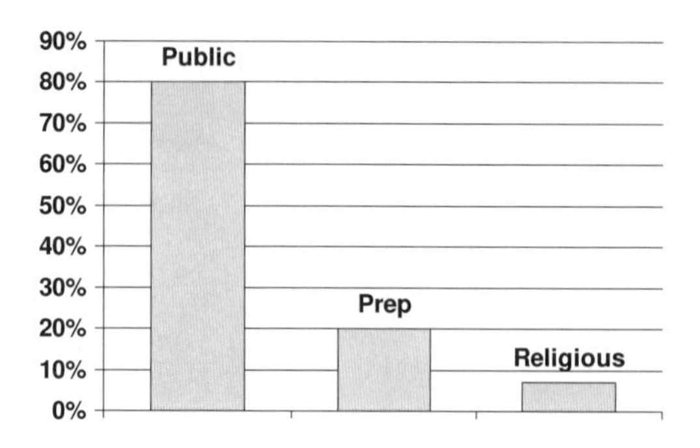

FIGURE 5. The type of high school that the CAARMS12 attendees went to.

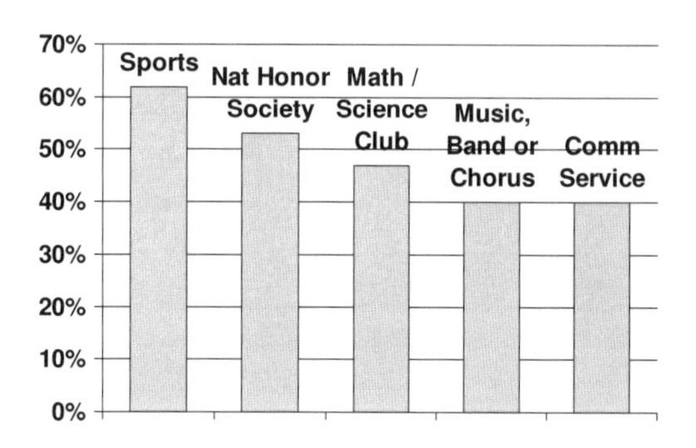

FIGURE 6. The type of extra-curriculum interests that the CAARMS12 attendees had in high school.

2.8. What type of colleges did attendees go to? The most popular university system for the participants at the conference were *historically Black colleges and universities* (HBCUs), see Figure 8. There were 42% hailing from such institutions, while 29% were from predominantly white private institutions and 29% were from predominantly white state institutions. These trends are quite surprising in some regards as the majority of black youth no longer attend HBCUs but predominantly white state institutions. One possible reason for the high numbers of Ph.D. students coming from HBCUs could be from the nurturing that may take place at an HBCU that may not occur at a predominantly white state or private institution.

Several researchers (e.g. Hilliard [**Hi**], 2003; Southern Education Foundation [**S**], 2005) have found that HBCUs have specific characteristics that explain their success in steering Black students through the *science, technology, engineering, and mathematics* (STEM) pipeline. They provide opportunities for students to excel, facilitate peer groups that support mathematics engagement, express high expectations for the success of black students, and provide intervention for students, when

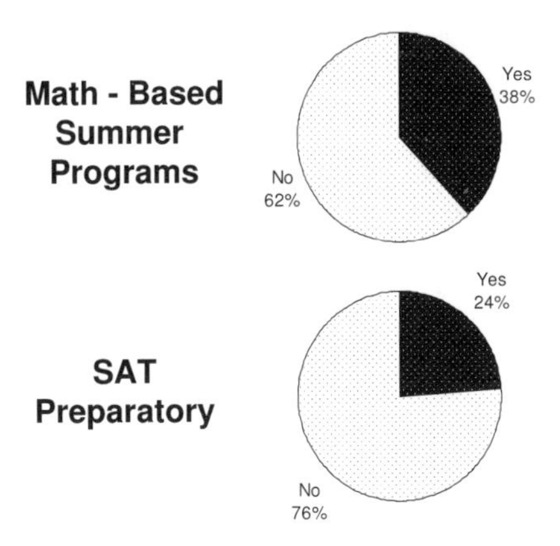

FIGURE 7. The type of involvement that the CAARMS12 attendees had in math or science based summer programs in high school.

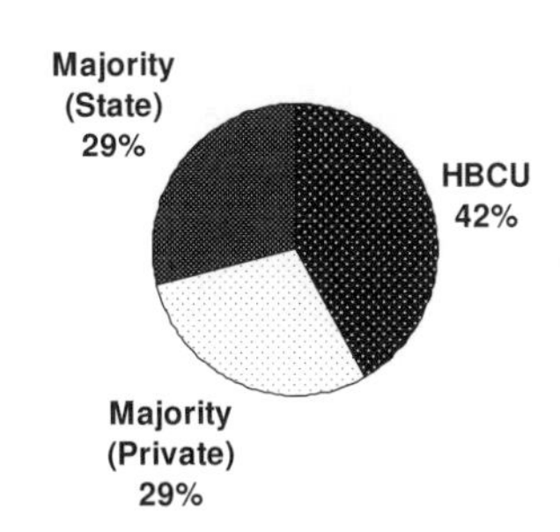

FIGURE 8. Survey questions about the type of extra-curriculum interests that the CAARMS12 attendees had in high school.

needed, early and often. In many ways, effective college programs for increasing STEM participation by students of color at predominantly white institutions have replicated the best practices of HBCUs.

2.9. What were participant's undergraduate majors? Sixty six percent of conference attendees majored in a mathematical science. Engineering as a major was a distant second, with 16%. Attendees also reported majors in science (9%) and "other" majors (9%), see Figure 9.

2.10. What awards did participants win for college and graduate work? This study showed the great importance of industrial and government sponsored scholarships for blacks seeking a Ph.D. in the mathematical sciences, see Figure 10. With 16% of participants having their undergraduate education partially

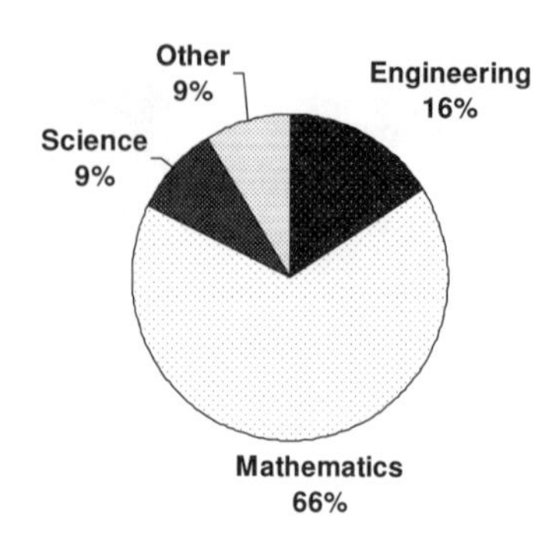

FIGURE 9. The college undergraduate majors of the CAARMS12 attendees.

funded by industrial college scholarship and 17% obtaining industrial graduate fellowships, the importance of such funding is evident. These monies are even more vital with the current rising costs of higher education compared to their expenses in earlier times. With 22% of attendees obtaining government college scholarships and 51% winning government graduate fellowships, government support was responsible for almost half the attendees at the conference. There have been recent cut backs in government spending on initiatives to help support blacks in higher education. This could have a negative impact on the black pipeline to the Ph.D. in mathematical fields unless funding for programs is restored in the near future.

2.11. Which degrees did participants already have? About 70% of participants to the conference had degrees in the mathematical sciences, see Figure 11. This is not surprising when considering that the conference was one based on these subjects. There were also significant numbers of participants who had engineering degrees (16%) as well.

2.12. In what fields do participants plan to pursue a Ph.D.? About 42% of participants will seek a Ph.D. in Pure Math while 23% will pursue one in Applied Math and 11% will strive for one in Operations Research (OR), see Figure 12. The numbers obtained from these three categories make up the overwhelming majority of those participants seeking a Ph.D. in the mathematical sciences at the conference.

3. Conclusion

This conference obtained survey data on half of the CAARMS12 participants who had obtained or were seeking a Ph.D. in the mathematical sciences. Our analysis showed that many individuals came from solidly middle or upper-middle class backgrounds but not exclusively. One notable characteristic that was garnered through the in-depth interviews were the high numbers of participants coming from two-parent households. Other factors seem to arise in these surveys that give a sense of what it takes to obtain a Ph.D. Factors such as a supportive family structure and significant involvement in extracurricular activities in high school such as sports, music or math/science based clubs.

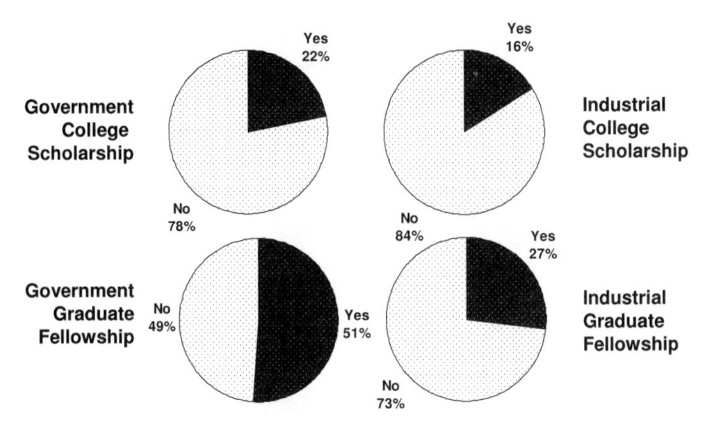

FIGURE 10. Survey question about college scholarships awarded to the CAARMS12 attendees.

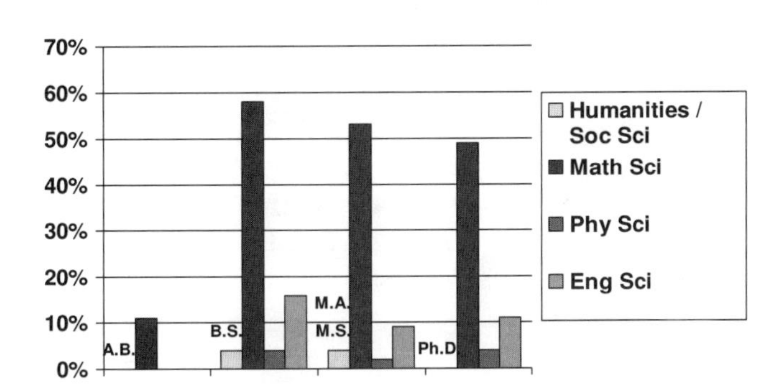

FIGURE 11. The undergraduate and graduate degrees that the CAARMS12 attendees currently have.

This exploratory study, even with its limitations, has revealed several potential avenues for further examination. First, more information should be gathered about the kinds of supplementary educational activities and formative childhood experiences that African American Ph.D. holders and candidates in the mathematical sciences have as children and adolescents. Second, additional probing should be done about the kinds of extracurricular activities that were available to students, and how other out of school activities-perhaps structured by parents-helped to augment school experiences of the respondents. Finally, given that similar numbers of African American male and female students are successful in gaining the Ph.D., while for other groups-including Africans-the results are skewed toward males, further study of the dynamics that contribute to this parity should be explored.

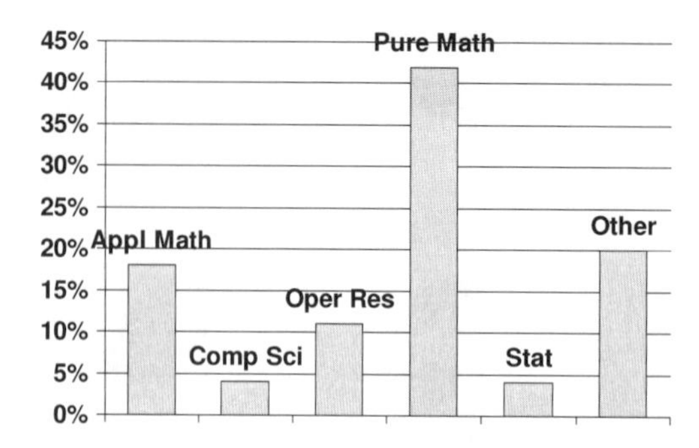

FIGURE 12. The technical fields for which the CAARMS12 attendees plan to (or already have) pursue a Ph.D.

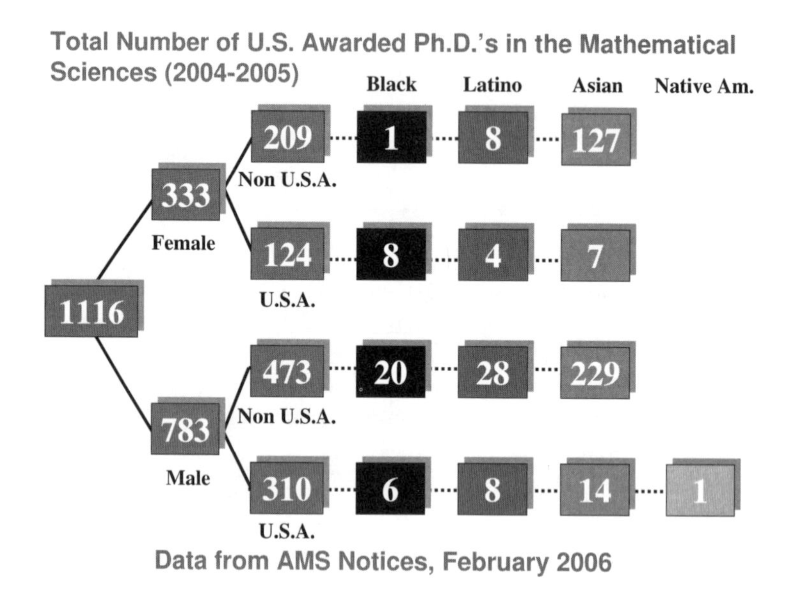

FIGURE 13. AMS data on the demographic breakdown of mathematics Ph.D.'s awarded in the U.S.A. during 2004-2005.

4. Appendix

The Conference for African American Researchers in the Mathematical Sciences (CAARMS) came into existence in 1995 to encourage, nurture, and promote African American researchers in the mathematical sciences, current and potential. We define the mathematical sciences in the broadest sense to include pure and applied mathematics, operations research, statistics, computer science and other technical fields where mathematics plays a major role. The conference also exists to complement and cooperate with other established groups that promote African-Americans in the mathematical sciences.

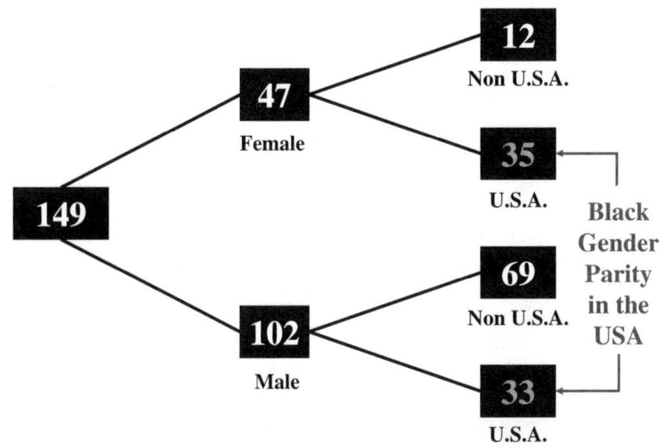

Data from American Mathematical Society (AMS) Notices

FIGURE 14. AMS data on the citizenship and gender breakdown of Black mathematics Ph.D.'s awarded in the U.S.A in the 21st Century.

The conference is primarily (but not exclusively) aimed at showcasing the research of African-Americans mathematicians. At the same time we have invited noted Hispanic mathematicians to speak at CAARMS such as Richard Tapia or Rice University, Rodrigo Bauelos of Purdue University and Carlos Castillo-Chavez of Arizona State University as well as Native American mathematicians such as Robert Megginson of the University of Michigan.

We have also invited mathematicians from the majority community to speak at CAARMS such as Steven Shreve of Carnegie Mellon University, Jerry King of Lehigh University, Robert Vanderbei of Princeton University, Margaret Wright of New York University, Ward Whitt of Columbia University, and Christopher Jones of the University of North Carolina at Chapel Hill and SAMSI. We have also invited speakers from the international community such as Carl Graham from the École Polytechnique in Paris.

The format of the conference is to have 10 researchers each give hour long talks about their research. Additional lectures may be given of an expository nature like a tutorial. At the end of each day, we have complementary events such as on Wednesday there is a graduate student poster session. on Thursday there is a conference banquet and a keynote speaker. Finally, on Friday we have a closing group discussion about careers in the mathematical sciences or future goals for CAARMS. One key feature of our conferences is that we do not have any parallel sessions. This encourages everyone to attend all the lectures and get exposed to different aspects of research in the mathematical sciences.

The conferences have been successful in attracting some of the best African American mathematical sciences graduate students in the country to CAARMS conferences. Since 1996, we have consistently had 20 graduate students participate each year for the CAARMS poster sessions. Over 70 African American graduate students from the previous CAARMS meetings have since gone on to receive a Ph.D. in the mathematical sciences. We have also had over 13 conference participants go on to become tenured university professors in the mathematical sciences.

References

[G] Gordon, E.W., Bridglall, B.L., and Meroe, A.S. (2005). Supplementary education: The hidden curriculum of high academic achievement. Lanham, MD: Rowman and Littlefield.

[He] Herzig, A.H. (2004). Becoming mathematicians: Women and students of color choosing and leaving doctoral mathematics. Review of Educational Research, 74 (2), 171-214.

[Hi] Hilliard, A. (1995). The maroon within us. Baltimore, MD:Black Classic Press.

[K] Kenschaft, P.C. (2005). Change is possible: Stories of women and minorities in mathematics. Providence, RI: American Mathematical Society.

[M] Massey, W. A. (2001). Mathematics is Four Dimensional. *Council for African American Researchers in the Mathematical Sciences: Volume III. Contemporary Mathematics*, editors Alfred Noel, Earl Barnes and Sonya A. F. Stephens. Volume 275, pp. 147–158.

[S] Southern Education Foundation (2005). Igniting potential: Historically black colleges and universities and science, technology, engineering, and mathematics. Atlanta, GA: Author.

[W6] Walker, E.N. (2006). Urban high school students' academic communities and their effects on mathematics success. American Educational Research Journal, 43 (1), 41-71.

[W8] Walker, E.N. (2008). Mathematical (role) models: How Black mathematicians mentor, teach, and practice. Paper presentation at the Annual Meeting of the American Educational Research Association, New York City, NY.

Department of Operations Research and Financial Engineering, Princeton University, Princeton, New Jersey 08544

E-mail address: wmassey@princeton.edu

Duke University School of Law, Durham, North Carolina 27708

E-mail address: derrick.raphael@law.duke.edu

Program in Mathematics Department of Mathematics, Science, and Technology Teachers College, Columbia University, New York, New York 10027

E-mail address: ewalker@exchange.tc.columbia.edu

Titles in This Series

445 **Laura De Carli and Mario Milman, Editors,** Interpolation theory and applications, 2007

444 **Joseph Rosenblatt, Alexander Stokolos, and Ahmed I. Zayed, Editors,** Topics in harmonic analysis and ergodic theory, 2007

443 **Joseph Stephen Verducci and Xiaotong Shen, Editors,** Prediction and discovery, 2007

442 **Yi-Zhi Huang and Kailash C Misra, Editors,** Lie algebras, vertex operator algbras and their applications, 2007

441 **Louis H. Kauffman, David E. Radford, and Fernando J. O. Souza, Editors,** Hopf algebras and generalizations, 2007

440 **Fernanda Botelho, Thomas Hagen, and James Jamison, Editors,** Fluids and waves, 2007

439 **Donatella Danielli, Editor,** Recent developments in nonlinear partial differential equations, 2007

438 **Marc Burger, Michael Farber, Robert Ghrist, and Daniel Koditschek, Editors,** Topology and robotics, 2007

437 **José C. Mourão, Joào P. Nunes, Roger Picken, and Jean-Claude Zambrini, Editors,** Prospects in mathematical physics, 2007

436 **Luchezar L. Avramov, Daniel Christensen, William G Dwyer, Michael A Mandell, and Brooke E Shipley, Editors,** Interactions between homotopy theory and algebra, 2007

435 **Krzysztof Jarosz, Editor,** Function spaces, 2007

434 **S. Paycha and B. Uribe, Editors,** Geometric and topological methods for quantum field theory, 2007

433 **Pavel Etingof, Shlomo Gelaki, and Steven Shnider, Editors,** Quantum groups, 2007

432 **Dick Canery, Jane Gilman, Juha Heinoren, and Howard Masur, Editors,** In the tradition of Ahlfors-Bers, IV, 2007

431 **Michael Batanin, Alexei Davydov, Michael Johnson, Stephen Lack, and Amnon Neeman, Editors,** Categories in algebra, geometry and mathematical physics, 2007

430 **Idris Assani, Editor,** Ergodic theory and related fields, 2007

429 **Gui-Qiang Chen, Elton Hsu, and Mark Pinsky, Editors,** Stochastic analysis and partial differential equations, 2007

428 **Estela A. Gavosto, Marianne K. Korten, Charles N. Moore, and Rodolfo H. Torres, Editors,** Harmonic analysis, partial differential equations, and related topics, 2007

427 **Anastasios Mallios and Marina Haralampidou, Editors,** Topological algebras and applications, 2007

426 **Fabio Ancona, Irena Lasiecka, Walter Littman, and Roberto Triggiani, Editors,** Control methods in PDE-dynamical systems, 2007

425 **Su Gao, Steve Jackson, and Yi Zhang, Editors,** Advances in Logic, 2007

424 **V. I. Burenko, T. Iwaniec, and S. K. Vodopyanov, Editors,** Analysis and geometry in their interaction, 2007

423 **Christos A. Athanasiadis, Victor V. Batyrev, Dimitrios I. Dais, Martin Henk, and Francisco Santos, Editors,** Algebraic and geometric combinatorics, 2007

422 **JongHae Keum and Shigeyuki Kondō, Editors,** Algebraic geometry, 2007

421 **Benjamin Fine, Anthony M. Gaglione, and Dennis Spellman, Editors,** Combinatorial group theory, discrete groups, and number theory, 2007